SAILING SHIPS
FROM PLASTIC KITS

SAILING SHIPS

FROM PLASTIC KITS

KERRY JANG

Seaforth
PUBLISHING

© Copyright © Kerry Jang 2024

First published in Great Britain in 2024 by
Seaforth Publishing
An imprint of Pen & Sword Books Ltd
George House, Beevor Street,
Barnsley S71 1HN

www.seaforthpublishing.com
Email info@seaforthpublishing.com

British Library Cataloguing in Publication Data
A CIP data record for this book is available from the British Library

ISBN 978-1-3990-7860-3 (Hardback)

ISBN 978-1-3990-7891-7 (ePub)

ISBN 978-1-3990-7892-4 (PDF)

All rights reserved. No part of this publication may be reproduced or transmitted in any form or by any means, electronic or mechanical, including photocopying, recording, or any information storage and retrieval system, without prior permission in writing of both the copyright owner and the above publisher.

The right of Kerry Jang to be identified as the author of this work has been asserted in accordance with the Copyright, Designs and Patents Act 1988

Pen & Sword Books Limited incorporates the imprints of Atlas, Archaeology, Aviation, Discovery, Family History, Fiction, History, Maritime, Military, Military Classics, Politics, Select, Transport, True Crime, Air World, Frontline Publishing, Leo Cooper, Remember When, Seaforth Publishing, The Praetorian Press, Wharncliffe Local History, Wharncliffe Transport, Wharncliffe True Crime and White Owl.

Designed and typeset by Neil Sayer.
Printed and bound in China by 1010 Printing International Ltd.

Contents

Preface   6

Acknowledgements   6

1 Sailing Ships in Plastic   7

2 The Model Maker's Challenge   16

3 Building Out-of-the-Box: *Victory*, 1765. Airfix 1/87 Scale   34

4 Beyond the Box: HMAV *Bounty*, 1787. Airfix 1/87 Scale   47

5 Scenic Models: Jim Baumann's *Mary Rose* and *Sir Winston Churchill*. Airfix 1/400 & Aoshima 1/350 Scales   68

6 Hybrid Modelling: Victor Yancovitch's *Le Soleil Royal*. Heller 1/100 Scale   84

7 Modelling Realism. David Dovidio's *Vasa*. Airfix 1/144 Scale   90

8 Reconstruction of a Ship: Marc LaGuardia's *Le Soleil Royal,* 1689. Heller 1/100 Scale   100

9 Life at Sea: Daniel Fischer's HMS *Victory*. Heller 1/100 Scale   114

Resources   127

Preface

This book is aimed squarely at plastic modellers to help you get the best from your new plastic sailing ship kit regardless of its subject, scale, or style. The tools, techniques, and products required are those familiar to plastic modellers and if you are a wooden ship modeller you may find them surprising. But there is much for wooden ship modellers to learn from plastic modellers and *vice-versa*. Suspend any disbelief and instead ponder how they may help you build a better model no matter the material. Every modeller has different aspirations, preferences, and artistic styles, so pick and choose the techniques that work for you.

Acknowledgments

I'd like to acknowledge the importance of sibling rivalry in completing this book. I would have suffered the wrath of my wife when some errant model paint on the back of my hand smeared a newly upholstered dining room chair. Thanks to the quick action of my son to deflect responsibility to his younger sister, and her acceptance of the blame knowing that she would only suffer my wife's admonishment to 'wash your hands next time', I will be able to keep building models long into the future.

Kerry Jang
Vancouver, Canada

1: Sailing Ships in Plastic

Amongst all of humankind's creations, the sailing ship ranks with the most beautiful and awe-inspiring. Prior to the plastic kit, model ship kits were made of wood. Hulls were blocks of wood, or in more expensive kits, pre-carved to a rough shape that was to be shaped by hand to the final dimensions printed on the scale plans. Deckhouses and decks were cut from sheet, and the masts and yards made from lengths of birch dowel. Being made from wood, finishing took a lot of effort, requiring several coats of talcum powder mixed with model aircraft dope to fill the wood's grain before painting. Items we know today as decals were printed designs on the instruction sheet to be coloured, cut out and pasted to the model's surface. It was traditional model making in the truest sense, and the models produced from these kits were often basic in detail and over-scale.

Plastic kits at a stroke rendered wooden kits obsolete by promising simplified assembly and a fidelity to detail that was not possible before. Although the first plastic kits were sometimes toy-like, the kits gave the enthusiast a base from which to create a more authentic model. Plastic moulding technology quickly improved and soon hulls were cast in a few highly detailed parts festooned with realistic planking detail, sumptuous decorations, and even a simulated wood grain finish. Parts in some kits were moulded in realistic colours, such as black for the hulls and guns, light brown for

1

Old and new Airfix. The *Cutty Sark* (approx. 1/775 scale, shown on the right) was amongst the first plastic sailing ship kits issued. The header art is inspiring, as is the box art to the new 1/400 *Mary Rose* released in 2014. The two kits share a common design philosophy with simplified hull mouldings, and yards and sails moulded as single units. Despite their small size, both kits are finely detailed and build into neat replicas.

the deck parts, medium brown for the masts and yards, and white for the ship's boats in a bid to reduce painting. Parts were simply clipped from the sprue and assembled with no more than plastic cement, leaving the builder with a ship of their dreams in mere days instead of years.

A Little History

It comes as no surprise that the earliest plastic kits were of sailing ships. Airfix led the way with a range of small-scale bagged kits starting with the *Golden Hind* (1952), *Cutty Sark* (1955), and *HMS Shannon* (1954) whose sales encouraged the company to expand the range. These kits were simple and of an indeterminate scale, but were quick to assemble and at a distance had the outlines recognisable as their namesakes. The American company Pyro followed suit and in the 1960s released a very popular range of kits including Christopher Columbus' *Nina*, *Pinta*, and *Santa Maria*, the whaler *Charles W Morgan*, and famous American warships like *Constitution*, and her adversaries such as the *Barbary Pirate Corsair*. Pyro's offerings were also a success and spurred other manufacturers to offer sailing ships kits, but as the competition grew, so did the range of ships, their scale, and their complexity, which was a boon to model shipwrights around the world.

Airfix again led the way and throughout the 1960s and 1970s released highly detailed scale kits of popular ships including HMS *Victory* (1/180), *Prince* (1/180), *Royal Sovereign* (1/168), Captain James Cook's *Endeavour* (1/120) and William Bligh's *Bounty* (1/87). Their last sailing ship was the French armed Indiaman *St Louis* (1/144) in 1975. These kits were designed using the best available research at the time and featured fine detail. The ship's sails had evolved and were no longer thick and heavy, moulded integrally to the yards, but now vacuformed out of thin styrene sheet to a much more scale thickness. In the late 1960s Heller of France burst on to the scene with some of the most magnificent models of sailing ships ever seen in plastic. Their focus on the French navy through the ages brought us a series of 1/200 scale kits of *Le Royal Louis*, *Le Superbe*, *Le Gladiateur*, to the giant 1/100 scale *Le Soleil Royal*, the 1/75

2

Some Pyro kits like the *Constitution* were quite accurate in shape and detail, whereas other named vessels such as the *Bonhomme Richard* did not remotely resemble the French Indiaman *Duc de Duras* from whom she was converted. Many of the Pyro kit designs are imaginary like the *Mayflower* pilgrim ship for which no contemporary illustrations or drawings exist, leaving the toolmakers to base their kit on generic ships from the Elizabethan age. Despite their shortcomings, because the kits bear the names of these famous vessels, their popularity has endured. Many of the Pyro kits are now sold under the Lindberg Line label.

SAILING SHIPS IN PLASTIC

3
A selection from the classic range of Airfix sailing ships. Highly detailed and easy to assemble into authentic models, they also provide a superb base for super detailing. The range is quite extensive and many are currently being re-released in their Vintage Classic range. The *Cutty Sark* and *Endeavour* are accurate models that are a good match to extant drawings, but the *Golden Hind* is based largely on conjecture. That the full-sized replica displayed on London's Southbank and the kit differ in so many ways only demonstrates that different designers trying to reconstruct the ship can have very different interpretations. This should not deter you from enjoying the kit if you simply accept it for what it is – a reconstruction of how *Golden Hind* (*Hinde*) may have appeared.

galley *Le Reale de France*, and 1/50 *Le Chebec*. The research that went into the kits was first rate being based on the plan sets published by the Association des Amis du Museé de la Marine. The range culminated in 1979 with the ultimate plastic kit of Nelson's *Victory* at 1/100. The standard of detail, scale, and research was unprecedented, and these kits have never been eclipsed by modern toolings. Heller also boldly released kits of non-military ships, though with similar French flair, like the schooner *l'Amphrodite* (1/150), fishing vessels like *Thonier* (tuna fishing boat) to 1/125, and the towering masts of the German 'P' windjammers *Pamir* and *Passat* to 1/150. A glance through the Heller catalogues

over the decades really captures the age of sail like no other plastic range.

Major American companies like Aurora soon jumped on the bandwagon, releasing kits of the USS *Farragut*, USS *Constitution*, the Canadian racing schooner *Bluenose*, and Charles's I flagship *Sovereign of the Seas*. But it was Revell in 1959 who brought sailing ships to the American market in a huge way with their 1/96 scale *Cutty Sark*, heralding a new range of large-scale kits including a *Spanish Galleon* and the USS *Constitution* to the same scale that quickly captured the imagination of young and seasoned modellers alike. To entice modellers further, these kits were sometimes offered with real cloth sails or pre-painted hulls in authentic colours. Even their smaller-scale kits of ships like the *Bounty* (1/110) and *Mayflower* (1/183) beckoned, with ready to use rigging blocks and plenty of scale figures in period clothing to populate the decks. Despite these innovations, Revell was notorious for reboxing their kits as a different ship to get the most out of their moulds. Their generic *Spanish Galleon* was re-released as a generic *English Man O War*, the *Cutty Sark* was released as her arch-rival *Thermopylae*, and then again as the Portuguese naval training ship *Pedro Nuñes*. The American Civil War's USS *Kearsarge* was re-released as her adversary *CSS Alabama;* the *Mayflower* as an *Elizabethan Galleon*; and the USS *Constitution* as her near sister the USS *United States*. At least these rebrandings had a passing resemblance to their new identities, with a few extra parts

4
Heller's ship kits remain amongst the very best you can get in plastic. The fineness of the mouldings, attention to detail, high part counts, and broad range of subjects are hallmarks of the range. Heller kits are not nearly as simplified as many other manufacturers and modellers have remarked that they can be more of a challenge to build, but do build into highly detailed replicas out-of-the box. For two of their most complex kits, *Le Soleil Royal* and *Victory*, Heller have created new instructions (available separately) correcting errors and expanding on the rigging for these kits. After finishing the model these booklets serve as good general references to rigging ships of the seventeenth and eighteenth centuries respectively.

5
Heller's HMS *Victory* box art. This superb kit is slated for re-release.

6

A range of kits in a wide variety of scales to suit your budget and display space. Shown are Airfix's *Victory* (1/180), *Cutty Sark* (1/130), Heller's *La Reale de France* (1/75), *La Belle Poule* (1/200) and *Royal Louis* (1/200), Aoshima's *Chinese Junk* (1/350), and Langton's multimedia resin, white metal and PE brass HMS *Juno* (1/300).

tooled so they could be passed off as the new ships - but Revell's marketing department did not stop there. Most infamously, their HMS *Bounty* was repackaged as Charles Darwin's HMS *Beagle*, the ship that carried the naturalist Charles Darwin to the Galapagos Islands where he developed the Theory of Evolution. In reality, the *Bounty* and *Beagle* looked nothing alike and their design was separated by over a century.

Revell is not alone in this practice, and perhaps the most egregious example is when one manufacturer's kit is released by another manufacturer as a completely different, and often fictional vessel. For example, Heller's *La Belle Poule*, a frigate from the 1850s famous for returning Napoléon I's ashes to France from Elba was released by Zvezda of Russia as the *Acheron*, a fictitious ship from the *Master and Commander* motion picture starring Russell Crowe. Lindberg in the United States is a prolific offender, creating a whole line of fanciful pirate ships by rebranding old kits out of Pyro's line. The Dutch ship *Gouda* has become the Lindberg *Flying Dutchman* – moulded in glow-in-the-dark plastic for added

7

Heller kits sport exquisite detail, as shown on these carved friezes from their *La Reale de France* kit. Carvings such as these highlight the advantage of plastic kits. To replicate these yourself would take years of carving experience and practice that would discourage most hobbyists from even starting. Underneath the parts is one of the perfectly printed banner, tabard, and flag sheets that will fly on the model when complete. This galley was rowed by convicts, and a model full of oarsmen at their benches would be quite a sight. Soft plastic wargame figures in 1/72 scale intended for Roman wargame ships might be a source, or for perhaps something from a 3D model range whose files can be modified would work well.

8
Revell's 1/96 *Thermopylae* can be built as a fully rigged clipper or a *barque* (a vessel where the fore and main masts are square rigged, but the mizzen is rigged fore and aft) depending on what time in the ship's history you wish to portray her. In this 1960 issue of the kit, the hull is moulded in green plastic and the copper bottom is pre-painted copper. The kit was originally released as *Cutty Sark*, with Revell ignoring the dimensional differences between the two ships. A few extra parts were included to ring in the changes. The *Cutty Sark* version remains a mainstay of Revell's catalogue to this day and builds into an impressive model out-of-the-box.

fun. The French galleon *La Couronne* became 'Sir Henry Morgan Pirate Ship'; and the list goes on. At least Revell's 1960 *Jolly Roger* (1/72) kit was based on something tangible – Disney's Peter Pan movie. *Caveat emptor*.

Across the globe, Japanese model companies also produced large numbers of kits. During the 1970s Imai released several large-scale kits of unique subjects not seen anywhere else such as the hybrid steam/sail line of battleship *Le Napoleon* (1/150), and the 1/180 scale paddle frigate *Susquehanna* (the American ship that opened Japan to the rest of the world), a *Chinese junk* (1/160), and Japanese sail training ships *Nippon Maru* and *Kaiwo Maru* to 1/100. Uniquely, Imai also released two sets of figures for the latter two kits to man the yards in salute or working the deck.

Interestingly, these plastic kits replaced the wooden kits of these ships offered by the same company. The Japanese have a penchant for offering the same subject in several scales to suit the space available in Japanese homes. For example, kits of the *Nippon Maru* are offered in 1/100, 1/150, 1/350, 1/200, 1/400, 1/900, 1/1000 by as many different manufacturers, and repeated again by reboxing them as her sister ship *Kaiwo Maru*. But be aware that not all *Nippon Maru's* are alike – in 1978 Arii released a 1/150 kit of the Shogun Toyotomi Hideyoshi's fourteenth-century galley that bore the same name. Such is the range of ship options available to the plastic modeller. Presently, Aoshima's small-scale (1/350) kits of modern sail training ships from maritime countries worldwide are very popular and make up into a neat collection. These small kits epitomise the small-scale genre with finely detailed and crisp accurate mouldings.

Since the 1980s releases of new kits have dropped off significantly. Most manufacturers have focused on re-releasing their classic kits or revamping them. There have been a few new releases, such as Revell's 2011 kit of *Vasa* (1/150) which is notable for incorporating the latest research from the *Vasamuseet* (Vasa Museum) in Stockholm. But not all new releases are an improvement. Revell's 1/225 HMS *Victory* (2016) is overly simplified and oddly detailed and their original *Victory* from 1958 is superior in virtually all respects. It is fair to say that the lack of new kits is made up for by originality. For example, Trumpeter of China released

SAILING SHIPS IN PLASTIC

9

Revell's (1/96) USS *Constitution* builds into a detailed replica out-of-the-box. All that was added to this model was a wooden deck overlay, some wooden rigging blocks to supplement the plastic kit's blocks, and cloth sails. The rest came from the kit box. Victor Yancovitch rigged all of the ship's guns with full breeching and training gear. The cloth sails were given *reef lines* from thread – short pieces of rope used to tie up the sails when they were being taken in (*shortening* sail) to slow the ship down, or if the wind is so strong that it might put undue stress on the masts and yards. All of the running rigging is properly belayed to the correct pin on the rails. (Photo used with permission)

10

Aoshima's *Susquehanna* was originally released by Imai of Japan. It is an impressive, if not simplified large scale kit. In Japan, all of the ships of Commodore Perry's fleet that forced Japan to open for trade are generically referred to as the 'black Ships'. This practice is carried over to the kit where the ship is boldly billed as a *kurofune* (黒船). Her actual name is only presented in English and in katakana (Japanese alphabet for foreign words) underneath.

a very intriguing large but indeterminate scale model of a *Ming Dynasty Treasure Ship*. Massive fleets of these ships commanded by the eunuch Zheng He (黒船 also transliterated as *Cheng Ho*) who led a fleet of these ships from China to the Middle East and across the Pacific in several expeditions between 1405 and 1433 searching for trade and treasure. At that time these expeditions showed that China was self-sufficient that led Emperor Zhu Di (朱棣) to cancel future voyages and adopt a general policy of isolationism that did not end until the end of Qing Dynasty in 1911. The treasure ship kit is based on museum models and drawings appearing in Chinese historical texts, but the remains of an actual ship have yet to be found.

A new trend for plastic models are resin kits and files (.stl) of ships that can be printed at home on a 3D printer. At this time the offerings are generally to a small scale (*eg*, 1/1200 to 1/700) and most are intended for naval games. But don't get the idea that a wargaming model is a simplified model. The reality is quite the opposite because 3D modelling permits the designer to incorporate exquisite detail into their products, and they take advantage of the fact that the 3D printing process creates a model by building it up in layers. This layering makes it is possible to create *open* gun ports with fully detailed guns sticking out through them in the smallest scales. Moreover, these models can also be scaled to different sizes to suit your needs. If you are able to print your own models, the files are quite inexpensive and whole fleets can be modelled. Imagine, printing off all the French and British ships that fought at the Glorious First of June and setting them is a seascape. How better to capture the glory and terror of naval combat.

The purpose of this book is to provide you with ideas and techniques to get the most of your model kit inspired by expert modellers from around the world. Each of the

11

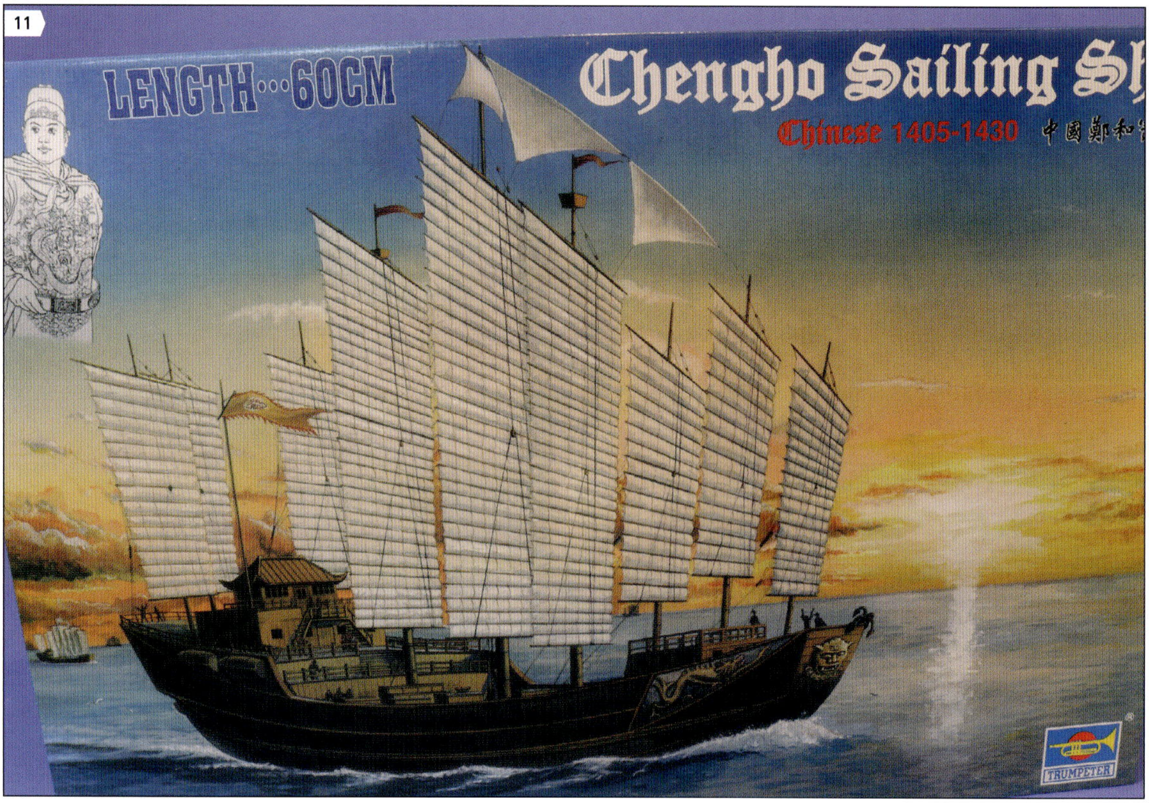
LENGTH…60CM
Chengho Sailing
Chinese 1405-1430
TRUMPETER

12

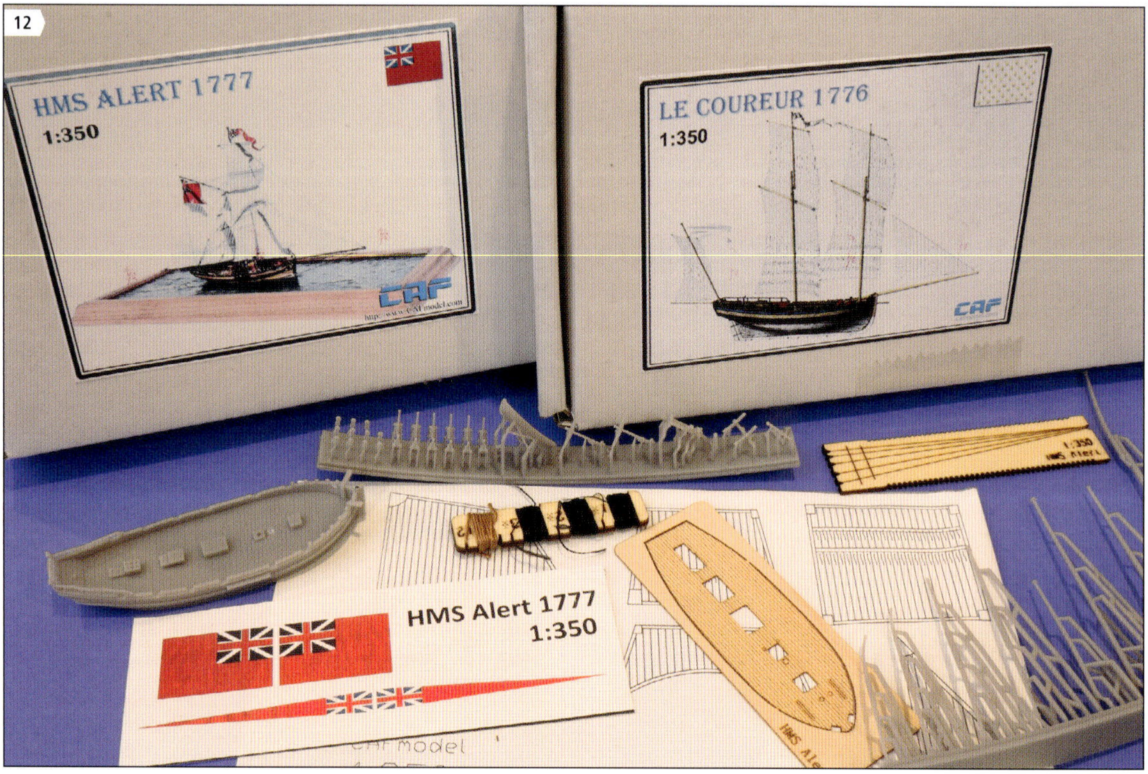
HMS ALERT 1777
1:350
CAF
LE COUREUR 1776
1:350
CAF
HMS Alert 1777
1:350

11

A treasure ship of the Ming Dynasty by Trumpeter Models in China to 1/250 scale. However, the width of the moulded planking detail suggests the kit's scale is more like 1/72. If so, the model would represent one of the smallest treasure ships. Chinese records give the size of the ships as 44 zhang (丈) which translates to over 140m (459ft) in length, while other historians have stated that Zheng He's largest ship was only about 70m (230ft) or less in length. To bring the ship more in line with 1/250 scale to represent the most famous versions of the ships, all of the planking should be filled and sanded smooth. Chinese records also suggest the ships were brightly painted and not the overall brown of the kit box, and so any planking lines in this scale would appear smooth under a layer of paint. Adding some scale figures would also give the viewer an appreciation of the size of the ships.

Anomalously, the kit's banners use the post-1949 simplified Chinese character 郑 for the surname Zheng (or Cheng), not the traditional version used in the Ming Dynasty. Several countries visited by Zheng He have dedicated museums to the mariner, and there is even an International Zheng He Society headquartered in Malacca City, Malaysia.

12

CAF Models of Shanghai have released two 1/350 3D printed resin kits of the British cutter *Alert* and the French lugger *Le Coureur*. These two ships faced off during the War of American Independence where *Le Coureur* was captured after a chase. The hulls, masts, yards, and fittings are supplied as 3D printed resin parts. The kit includes a wood veneer deck that has been fully detailed by laser etching. A laser cut loom in wood is included for weaving the shrouds and ratlines from thread, with sails and accurate flags in paper. These models are highly detailed and can be used as part of a diorama or as wargame pieces. The fine detail possible from 3D printing puts larger conventionally injection moulded styrene kits to shame.

13

If you own a resin 3D printer, Henry Turner designs are highly accurate and feature bold but in-scale details making them easy to paint. The photo show the French 80-gun ship *Le Tonnant* fresh off my 3D printer's building plate (top). The model is designed to 1/700 scale but was rescaled to 1/1200 scale (centre) without loss of any detail. The raw prints in grey resin show that even when scaled down to 1/1200, you can still clearly see the cannon and carriages in a fully hollowed ship's waist.

The 1/700 model in the foreground was printed and painted in the same afternoon. After the model was primed, the hull sides were airbrushed yellow ochre and the decks in grey-tan acrylic paints. With a fine brush, all the remaining details were picked out using acrylic paints from Vallejo. Each colour was next highlighted with lightened versions of the base colour, and when dry the whole model received a black-brown wash made from artist's oils dissolved in odourless thinner (eg, Gamblin's Gamesol). The wash defines the detail and tones down the paint colours. A clear coat of gloss acrylic varnish deepens the finish, and a coat of satin varnish provides the final sheen. The paint scheme used on *Le Tonnant* came from the colour illustrations in Jean Boudriot's 1986 book, *Le Vaisseau de 74 Canons*.

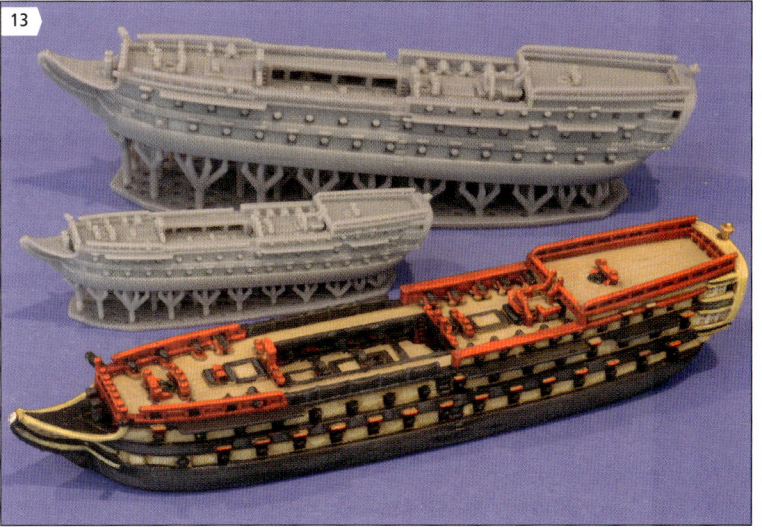

following chapters will feature a build of a plastic kit, beginning with some straightforward modifications and progressing to some very extensive rebuilding incorporating the latest research to create the most accurate model possible. Also featured are the techniques to realistically set a model into a seascape, and models that portray life at sea so authentically that it can be felt and almost experienced by the viewer. From these builds, you the modeller, can pick and choose what you like for your own model, and at the same time be inspired by what is possible.

2: The Model Maker's Challenge

Models of all subjects have their challenges to make them more authentic. For example, on armoured fighting vehicles getting the tracks to sag realistically, or how to add missing details like seatbelts to a model aircraft cockpit. Sailing ships are no different and when you open up the box for the first time you will get a good idea of what the kit's strengths and weaknesses are. However, the parts inside a sailing ship box look very different from an aircraft kit. In an aircraft kit, instantly recognisable are the parts for the fuselage and wings, and a bunch of little parts ('fiddly bits') for details. Opening a sailing ship kit the hull halves, and decks parts are recognisable, but the majority of the kit parts are these fiddly bits. There are many oddly shaped pieces of plastic making up the details for things you might not be familiar with, like *binnacles*, *bitts*, *channels*, *chains*, *sheer poles*, *dolphin strikers*, *knees*, *cheeks*, *hearts*, *catheads*, and *trailboards*. These items are all central aspects of a sailing ship and on a model each is itself made up of several individual parts. Indeed, for those new to sailing ships the first task it to learn the terminology, a peculiar vocabulary that has been developed over the centuries. Not only are the parts of a ship described and defined by specific names, but actions like *bending, reeving,* and *seizing*, not to mention sailing terms like *tacking* and *luffing* for example, are part of this specialised language. The best way to learn this language is to use it. The more it's used the more it will make absolute sense and you will come to realise what a handy language it is to describe and use all those fiddly bits. This book will provide some key definitions when they are first introduced and then use them throughout.

With a new kit, here are a few questions to ask yourself when you are gazing down at all the plastic goodness sitting on your lap. Do the hull parts capture the shape of the real ship? Do the parts feature a simulated wood grain finish? Do I even like woodgrain finish? Are the moulded hull planks to scale and are they represented by raised lines or engraved detail? If the ship had her bottom sheathed in copper, do the hull parts carry this detail, and if so, how are the copper plates represented? Do the moulded plates overlap each other slightly like the real ship? Do the plates carry nail head detail? If not, how do I model a copper bottom? Are the gunports recessed squares or are they pierced all the way through the hull? Then there is the ship's decoration – are the carvings authentic? Do the figures look human or were they sculpted with the proportions of an orc or troll?

Moving on to the decks, are the deck planks to scale and does the end of the plank (*butt*) show the proper *shift* so that their ends do not fall adjacent to one another? Are the ends of the planks at the bow and stern of the ship properly *joggled*? That is, as the shape of the deck at the bow and stern narrows, the deck planks must be narrowed and cut off square to fit. In some kits this important detail is missed and the plank ends are just cut off diagonally. Moreover, is the deck detailed with a moulded *margin plank* (a plank forming a border for the wooden decking), and are the deck planks cut into the margin plank properly? How do the deck hatches and gratings appear? Are the grating holes over-scale so that a scale *matelot's* (French for sailor) foot can get caught and break his ankles?

Sailing ships are very complex and kits often simplify or omit parts of the ship altogether. For example, deck fittings such as *bitts* (a pair of posts on the deck of a ship to fasten mooring or rigging lines) or belaying pins and their racks are often omitted to simplify rigging. Instead of tying a line to a specific belaying pin (each pin was assigned to a particular line so the sailor knew exactly where to find it) or bitt, the

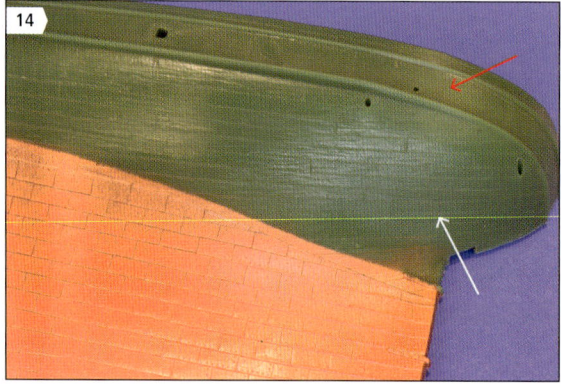

14

The clipper *Thermopylae* was *composite built*, meaning that the hull frames that shaped her hull were made of iron and planked over with wood, and that the upper *bulwarks* (ship's sides) were iron sheets riveted to the frames. Revell's kit accurately represents a composite built hull by carrying carry fine raised wood planking detail on the lower hull (white arrow) and bulwarks moulded as smooth with rivet detail (red arrow). The ship's copper bottom is authentically rendered with nail head detail visible under the factory applied copper paint in this 1960 issue of the kit.

THE MODEL MAKER'S CHALLENGE

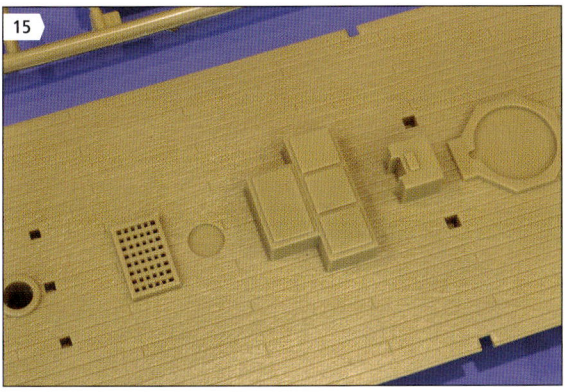

available books on popular vessels that are perfect for detailing a model. These books provide scale drawings of hull and deck details so you can easily identify missing fittings and the information to make or modify kit items you find unacceptable. Some specialist books illustrate the structural details of almost every piece of timber used to make a ship's frame, her beams, planking, and *knees* (wooden brackets used to support large timbers) and how they were joined so you can understand how each timber and fitting are intimately interconnected.

Now have a look at the masts and yards. Are mast parts

15
Aoshima *Susquehanna*'s main deck moulding sports engraved planking lines with a wood grain finish. In 1/150 scale the deck planks are too wide, as are the holes in the grating.

kit will instruct you to tie the line to any convenient railing instead (or simply omit the rigging line altogether). Small details like *binnacles* (the wooden cabinet that houses the ship's compass) and *hammock cranes* (iron stanchions rigged to hold the crew's rolled up hammocks) are rarely reproduced. If the items are not missing, then they are either heavily moulded, the wrong shape, or poorly detailed. The only remedy for this is to spend some time doing research. Ship modellers are fortunate in that there are several readily

16
The plastic deck from Airfix's *Bounty* kit (bottom). The deck planks are slightly too wide for the scale, but at the bow they are properly joggled (yellow arrow) into the margin plank (white arrow). However, the mouldings do not feature any butt detail suggesting the deck was planked with full length boards. If the kit parts are used, after painting the deck butts could be drawn in with a pencil, as could any *trennals* (wooden dowel or peg used to fasten pieces of wood together).

At the top is a laser cut and etched oak veneer deck by HiSModels. This deck was designed using authoritative references and features correctly scaled planking, greater joggling, the correct pattern of the butts (red arrow), and laser etched dots representing trennals.

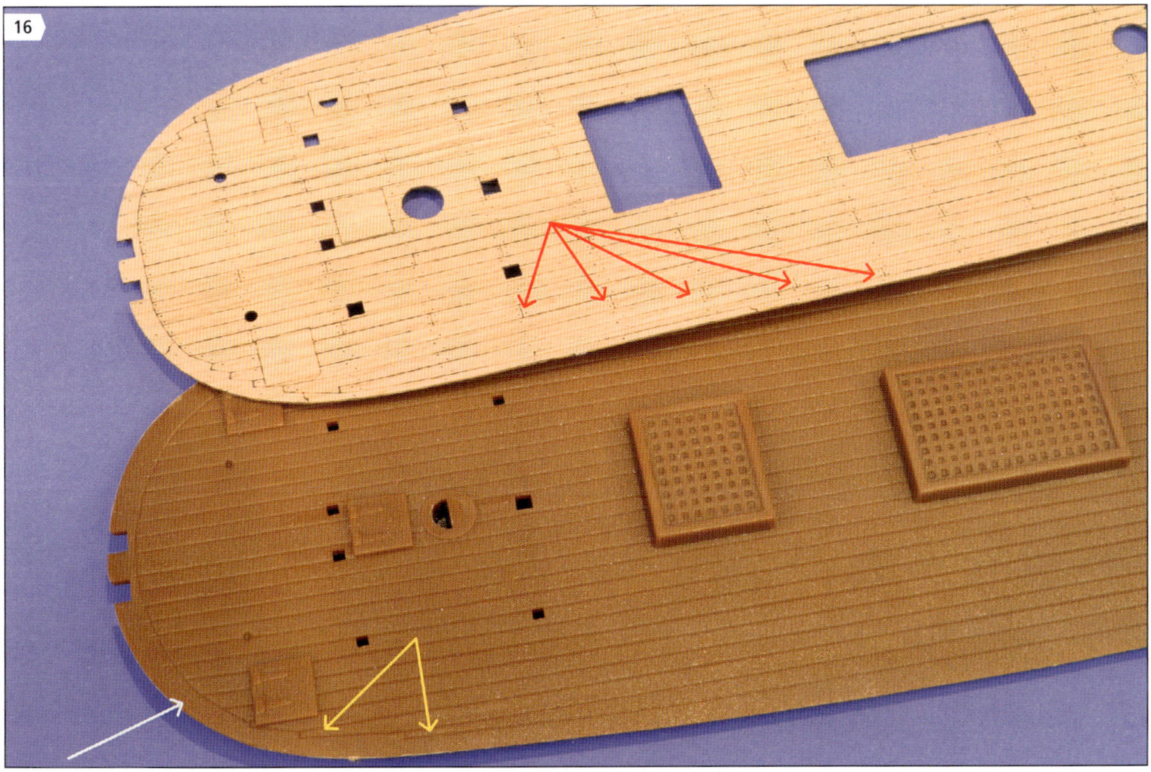

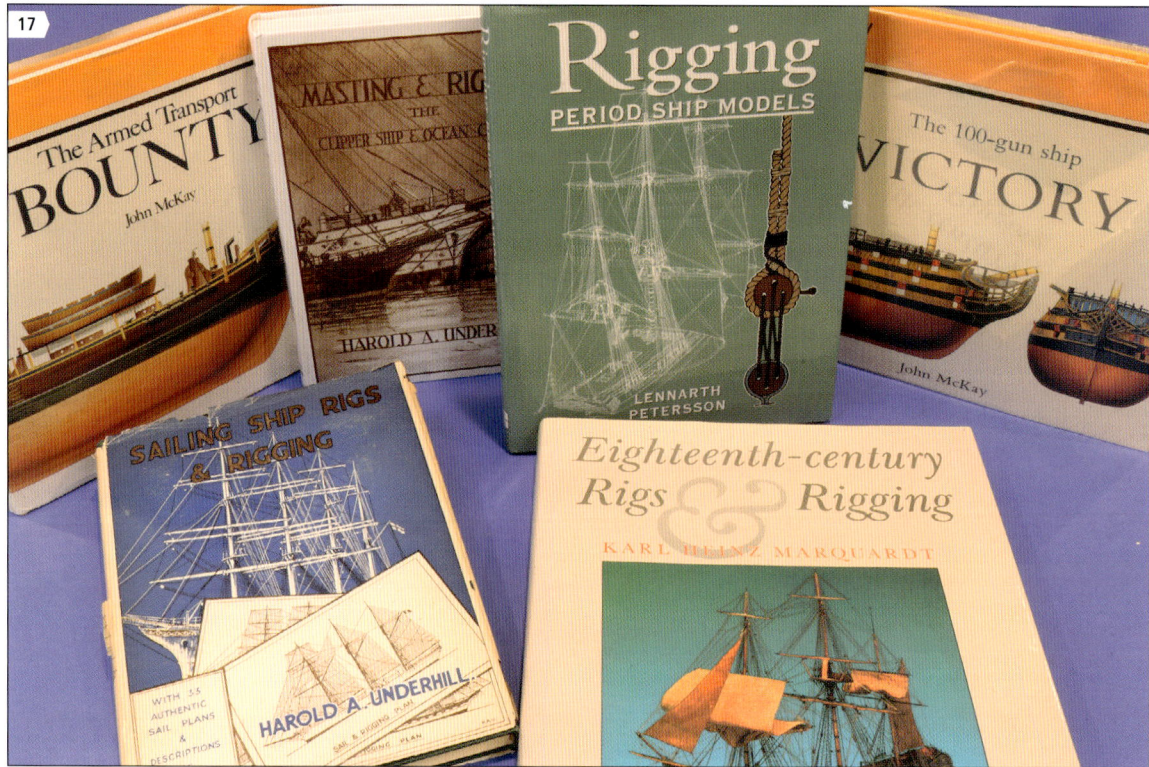

17

A selection of popular references for the sailing ship modeller in plastic or wood. Books like John McKay's *Bounty* and *Victory* contain pages of scale plans of those ships including full rigging plans and the dimensions of all the masts and yards. The books by Harold Underhill are excellent general references to rigging merchant ships. They were first published in the 1930s but remain the best references and are still in print. Lennarth Petersson's *Rigging Period Ship Models* is an essential reference with step-by-step drawings showing the run of each line and the order they are set up on a square rigged ship. For a ship that has no square sails, his book *Rigging Fore and Aft Period Craft* is the best reference for model making. Karl Heinz Marquardt's *Eighteenth Century Rigs and Rigging* is good general reference illustrating different types of sailing rigs for ships of all sizes and types, and well as illustrating national differences in rigging practice. This selection of books just scratches the surface of what is available. There are entire volumes devoted to the arming and fitting of ships, rigs and rigging, and seamanship as well.

straight? Each mast is typically composed of three, sometimes four sections known as the *lower*, *top*, *topgallant*, and possibly a fourth mast called the *royal*. Do the kit parts properly capture the narrowing taper of each of the masts the higher up they go? Or, as on some kits, is each one of the mast sections moulded to a constant shape and diameter for strength? When plastic is moulded into a mast or yard part, the longer the length of the part the more flexible it becomes. This problem is exacerbated if the mast or yard is of a slim cross-section. This has implications for whether or not the kit parts will be able to remain straight when the strain of the ship's rigging is placed on them. Would the kit's slenderest masts and yards (*eg*, topgallants and royals) benefit from being remade from wood or brass rod for strength?

Remaking the masts and yards from a stiffer material is not that difficult. The easiest material is to use wood dowel of similar proportions to the part being replicated. Tapering is done with sandpaper. The dowel is laid on the workbench and a sanding bock is used to work in the tapers. The trick is to keep rotating the wood as you sand so the taper is even around the entire circumference. For the slenderest masts and yards, if fine wooden dowel is not to be found, a bamboo kebab skewer works well. They are easily sanded to the correct shape with sandpaper, and given the grain structure of bamboo, the skewers can be split to roughly cut in the taper with a knife before smoothing off with sandpaper. Bamboo has an amazing strength to weight ratio, and in cities like Hong Kong it is still used for construction scaffolding around the tallest skyscrapers. Brass rod is best reserved for the finest masts and yards. Tapering brass takes a lot of work because it is so hard that is requires files and a lot of elbow grease to shape it. Some modellers have advocated chucking the brass in an electric drill. The brass rod is held in a piece of folded sandpaper and spun. The sandpaper is

moved up and down the brass rod to work in the taper. This sounds good in theory but it is fraught with danger and I do not recommend it at all. The speed of the drill is difficult to control with one hand and the brass rod held in the other can easily bend caused by the centripetal force and whip around causing grave injury. The heat generated by the sandpaper on brass rod is substantial as well and can cause burns. Use wood where possible, and a straight length of brass rod used as is. In smaller scales the fact that the rod is untapered is not noticeable. For larger models, just use shaped bamboo skewers.

Masts and yards have a lot of detail so you have to ask yourself what kind of detailing is present on these parts. For example, some kits omit important features like the mast *bands* (metal or rope hoops to strengthen the mast) or the mast *boot* (a collar that seals any gaps between the mast and the deck to prevent the ingress of water)? Are the different mast sections moulded together as a single unit complete with moulded *caps* (a thick block of wood used to confine two masts together) or are they assembled from separate parts using caps or *trestletrees* (timber crosspieces that join the topgallant to the upper mast, see Figure 92). When moulded as a single unit the masts are usually out of scale and might benefit from cutting them apart, remaking them from shaped wood dowel or brass rod and joining them together prototypically. A major feature of the masts are the platforms (*top* or *fighting top*) at the head of the lower and top masts. Do these tops carry detail or are they moulded plain? Often

18

Many books available for modellers focus on the British practices that were adopted by several other navies like the United States Navy, for example. For modellers of French and other continental naval powers whom they influenced, the books published by ANCRE are the best references available. There are volumes devoted to specific ships, and shown here is a massive four-volume set of scale drawings of every part of a French 74-gun ship. To the left is a volume devoted to the history of the ships of Louis XIV's navy. Many of the books are now published in English.

important details are omitted, such as the railings (or *barrier*, to keep the crew from falling off them), planking detail, and ship's lanterns.

The same question can be asked of the plastic yards. Are they robust enough to take the strain of rigging and hang sail? Are *stunsail booms* (thin spars attached to the outer ends of a yard to carry additional sail) present? These sails and booms are important to have on ships like *HMS Victory* if she is being portrayed at the Battle of Trafalgar. The winds were very light that day and all possible sail was set to move Lord Nelson's columns into the line of French and Spanish ships. This little detail also highlights the importance of research on your ship. Just as important as the structural and technical details of the ship is knowing *how* she was sailed, *when* she carried a particular set of sail, *which* ship's boats

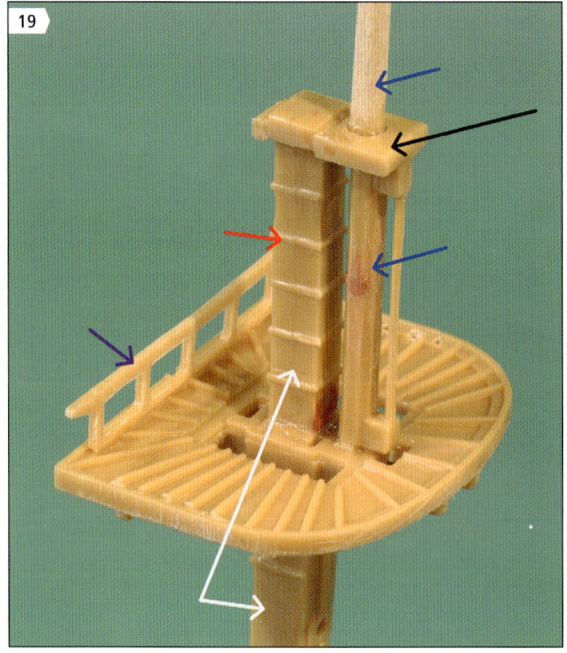

19

Airfix's 1/180 HMS *Victory* kit masts are well detailed and can be used out-of-the-box. The lower mast (white arrows) has moulded mast band details (red arrow), the mast top shows planking detail both on its upper and lower faces, and sports a guardrail (purple arrow) at the rear. The mast cap (black arrow) is moulded as separate parts so the upper and lower masts can be slid through and joined in an authentic fashion. The upper mast is a composite (blue arrows) of the kit part and a replacement made from a tapered hardwood dowel. The replacement was deemed necessary because the plastic part on my kit was moulded in a soft styrene and was prone to bending. Depending on when your kit was manufactured, a softer or harder styrene could have been used. This can be tested by trying to bend the part between your fingers and this test will give you an idea of how much strain they can take when rigging. In some models I have built, the masts were stiff enough to take the strain of the type of rigging I intended to use, and other times meant replacing a kit part, or changing the way I rigged the ship. The kit part was used as a pattern for the replacement. A hole was drilled into the cap and the new one glued in place.

20

The mast top after painting and highlighting with washes and drybrushing to really bring out the detail.

were carried, and *what* guns were present at a particular battle or point in time.

When it comes to the ship's rigging, I think it is safe to say that all plastic kits simplify rigging in order to streamline the construction of the model. This simplification takes on many obvious and some more subtle forms. For example, larger scale kits often provide rigging blocks but they are not really shaped like real blocks. Even if they are, do the blocks come in different sizes to handle different sizes of rope or are all one size? Sometimes the blocks are provided with a moulded ring ready for you to tie thread to, or they might be like real blocks requiring them to be *stropped* with thread – tying thread around the body of the block to form a ring (*becket*). A common simplification is to leave whole sets of lines off the ship or to instruct you to simply glue a piece of thread from one point to another. Sometimes why these two points were chosen has no practical basis in reality. At other times you are instructed to tie thread from one yardarm to the mast, with no indication that the line also should be run down to the deck so the sailor can handle it. The missing lines are not obvious to someone who is new to sailing ships, and the kit designers are clever enough to have you fit enough lines to give the impression of a fully rigged ship – even if the lines fitted are nonsensical. A dedicated model shipwright will want to rig virtually every line that existed on the ship, run them through the correct type of block and belay them to the correct spot. If opting for simplified rigging that leaves off some lines, then that same dedicated model shipwright will ensure that the lines that are fitted actually serve a purpose, are in the correct place, and most importantly, are clear in their function.

THE CHALLENGE OF PAINT FINISHING

Once you have come up with a list of what you will keep from the kit and what you want to change, your mind moves

THE MODEL MAKER'S CHALLENGE

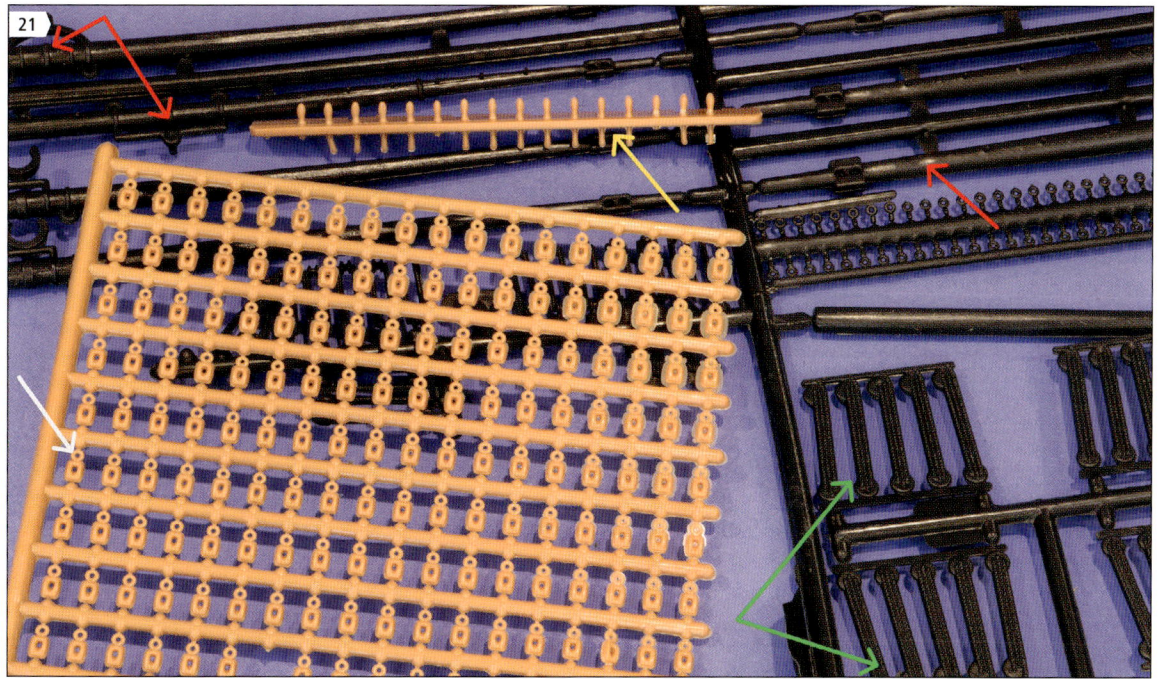

21

Many kits provide ready stropped rigging blocks where the modeller just has to tie a line to the moulded ring (*eye* or *becket*) at the top (white arrow). These parts come from Revell's 1/96 *Cutty Sark*. Also shown at the top of the photo are some of the kit's yards (red arrows). They are well moulded, robust, and carry fittings that are appropriate for a clipper of the 1870s. A belaying pinrail with moulded pins is also shown (yellow arrow). Although the part is well detailed, the bottom of each pin is quite fragile to the point where a few have snapped off. The bottom of the pins could be cut away and a hole drilled in its place. A short length of brass rod could be inserted into the hole and trimmed to length. This would provide a lot more strength to hold the rigging whilst keeping the well moulded pin top details intact. To the right, are integrally moulded deadeye and lanyard parts to simplify rigging (green arrows). For a modeller new to sailing ships the kit parts will familiarise you with ship's rigging, but as one gains experience they are best replaced with actual deadeyes and rigged like the actual ship for greater realism.

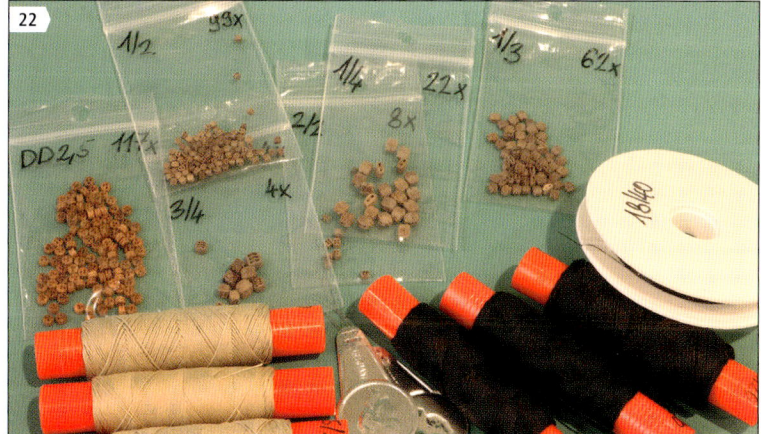

22

HiSModels of the Czech Republic (now known as Czechia) produce a number of rigging sets designed to fit specific kits of ships. The sets provide genuine wooden rigging blocks, deadeyes, and rigging thread in black (for tarred line) and tan (for untarred line) in several weights to better represent the different types of rope used on a ship. The wooden blocks are milled by laser and show all the proper sheave detail and scored around their circumference to take thread stropping. The shape of some blocks used by Continental navies differ from those used in the British navy and HiSModels offer accurate styles of blocks to reflect this.

Some of the blocks are so tiny that a sewing needle threader is included to help you get the lines through the miniscule holes! Dipping the end of the thread in CA creates a stiff 'needle' that easily allows the line to pass through the holes. ScaleWarship from the UK offer economical sets of realistic 3D printed rigging blocks in different scales, and Bluejacket Shipcrafters in the United States offer an extensive line cast in pewter.

to how to finish the model. Plastic kits are designed to wear an *authentic* finish in that the model is meant to look like the actual ship itself. This is quite different from wooden kits that are often designed to be *decorative*. In wood kits, little paint is used, instead relying on different species of wood to replicate the ship's colour scheme. The wood is usually given a coat of tung oil to bring out the colour of each species. Metal items like cannon are left in a polished brass, as are many smaller items like eyebolts, rings and hooks. The model itself may be 100% accurate in shape, structure and details, but are finished to highlight the beauty of the timber and metals used to make it.

When it comes to an authentic finish, we must tread carefully when we say that the model is finished 'like the real ship'. Many of our supposedly authentic finishes are best guesses. This is because detailed records of how a ship appeared were not kept, and only vague clues like 'she was bright finished' (meaning that her woodwork was not painted but varnished) is all that exists. Even famous ships like Bligh's *Bounty* have no official record of how she was painted, and the many models built of her are based on the sailing replica ships made for the movies and sport a scheme a prop maker dreamed up that would look good on camera. Models of ships that are housed in museums are often finished in the artistic style of the time as opposed to displaying an accurate representation of the actual colours. For example, it was long thought that *Vasa*'s carvings were gilt and mounted on a blue background akin to the colours of the Swedish flag. Once the ship was raised from the mud, it was found that the background was red, with all of her carvings fully painted in a multitude of colours. Some kits' paint instructions are based on research available at the time or on the colours of the museum ships (*eg, Cutty Sark,* HMS *Victory,* USS *Constitution*) but these colours often represent no more than the state of knowledge when they were restored, and use modern paints that are at best approximate matches to the period colours of the ship. Ships like *Victory* have been finished in a range of yellow paint shades during her many repairs ranging from a dark ochre to a bright yellow finish simply because the restorers used the closest shade they could find from commercial paint suppliers. Then, there are some finishing schemes that are just imaginary and fanciful because the actual ship has never been found (*eg, Nina, Pinta, Santa Maria*).

Another challenge is painting plastic to look like natural weathered wood. There are many paint techniques used by modellers to give a natural wood grain effect like those used on models of Great War aircraft fuselages, struts, and airscrews. A popular method is to undercoat the item with a light tan paint and brush over a coat of clear orange paint, or artist's oil colour such as raw umber streaked across the surface. One method gaining popularity takes advantage of the new ranges of highly pigmented inks (*eg*, Citadel Contrast Paint or Army Painter Speed Paints) used by wargamers to shade their miniatures. The parts are undercoated with off-white paint and these paints are simply brushed on. These semi-transparent paints seep into the recesses of the part to shade details and at the same time slightly flow off the top of parts to create highlights. They are simple to use and come in a wide range of browns and tans allowing almost any shade and species of wood to be recreated.

But what colour is brown? It really depends on the part you are painting as wood. Kit instructions will often tell you to paint a deck a generic 'light brown' or 'tan' colour. In reality, most decks are an off-white greyish colour caused by constant *holystoning* (scrubbing) by the crews and the effects of the sun bleaching the wood. Similarly, would a generic brown be good enough for the hull? Early wooden ships had their hulls *payed* (varnished) in a mixture of rosin, turpentine, and linseed oil that would produce a honey-toned finish. That may be the case when first applied, but that colour will shift after several coats of varnish. Moreover, because the varnish used at the time had a prolonged drying time it would have picked up dust, dirt, soot and other airborne contaminants that would most definitely cause an additional colour shift.

Many warships were painted in a scheme of ochre and black, but is ochre really yellow? Was the ochre of the British Navy the same as that used in the French Navy? In the days of sailing ships there were no national standard BS, FS or RAL paint colours dictated by the navy yards. Often the best information held in naval yard orders may direct the boatswain to add zinc oxide (white) to the ochre in in a 3:1 ratio to provide a 'pleasing finish'. Letters or accounts from ship's captains yield some tantalising hints, such as a British captain remarking that the ochre of their ships gave a dark and dingy appearance in contrast to the French whose ochre was lighter and brighter in colour. It's not even clear if ochre was yellow. For example, the HMS Victory Museum found that *Victory*'s well-known yellow ochre stripes really had a distinct *pink* undertone in 1805. What this all means is that when you decide on a colour for a sailing ship you must remember that paint was a natural product and depending on where the ingredients were mined, how it was thinned down by the crew, and how it was applied will yield a wide variation in colour. It is no use getting pedantic about it. Paint your model in the colour you think is most probable and if anyone says otherwise, ask them to provide you proof!

THE CHALLENGE OF COPPERING

The last major challenge is how to realistically portray a ship's copper bottom. On large-scale models, actual copper plates can be laid like the actual ship. This is a very common practice on wooden model ships. These copper plates can be used on plastic models as long as the scales are about the same. Most copper plates are stamped with a pattern of nail head detail that looks like large pimples. This detail can be hammered flat, leaving a much more realistic flat nail head

23

On Heller's 1/200 *La Belle Poule* the moulded copper plating detail on the lower hull was incomplete so the existing detail was sanded off. Any imperfections were filled, yielding a smooth hull ready for coppering. Self-adhesive copper tape (often found at stained glass window suppliers) was cut into rectangles and applied to the hull in brick-like fashion. Each row of copper was applied in prototypical fashion with the proper overlaps between each plate. It is a bit of an art to laying copper in that each *strake* (row) will take its own path up towards the waterline caused by the curve of the hull. The trick is not to force the rows to lie in any way but to let the row take the path it wants to. The resulting gaps between the rows are filled with copper plates later.

If you choose to copper your model, do some research before starting. The pattern of the copper on naval ships was different from that of merchant ships. On naval ships the coppering started at the keel and worked upwards. On merchant ships several (typically 6 or 7) parallel rows were laid along the waterline down towards the keel. At the same time, several parallel rows were laid along the keel and worked up towards the waterline. The areas between the upper and bottom rows (called the *goring strakes*) were filled in last, shaping the copper plates at the beginning and end of each strake to allow parallel rows to be laid to fill the space. This method was more economical in that fewer plates needed to be shaped with attendant wastage of copper, an important consideration in commercial ventures. Drawings will show the copper's pattern, size, manner in which each plate was nailed to the hull, and where each strake goes.

The photo also shows *La Belle Poule's* new *bowsprit* (a large spar projecting over the prow of the ship to provide additional sail – white arrow) made from kit parts with a *jibboom* (red arrow) and *flying jibboom* (attached to the end of the bowsprit though the cap – yellow arrow), *martingale* (in place beneath the bowsprit to attach rigging to stiffen the bowsprit – green arrow), and *jackstaff* (vertical rod just behind the cap where a jack – a type of flag – is flown when the ship is in port) made from brass rod (purple arrow). A new *beakhead* deck (a small deck set into the ship's bow – brown arrow) cut from scribed basswood sheet will replace the plastic part. A scribed basswood main deck will also replace all of the kit decks, using them as a pattern so they perfectly fit into the hull halves. If using scribed basswood for a new deck, the deck plank seams can be drawn in with a sharp pencil down each of the scribed grooves. Plank butt and trennal detail can also be drawn in. Scribed basswood decks are very bright in colour but can be weathered with thin washes of brown and black, and grey washes. These washes must not be made from acrylic paint thinned with water – or even their proprietary thinners which contain a high proportion of water – which will soak into the wood and cause warping. Instead, the washes must be made using enamel or artist's oils thinned in a petroleum based thinner such as white spirit, Varsol, or odourless thinners.

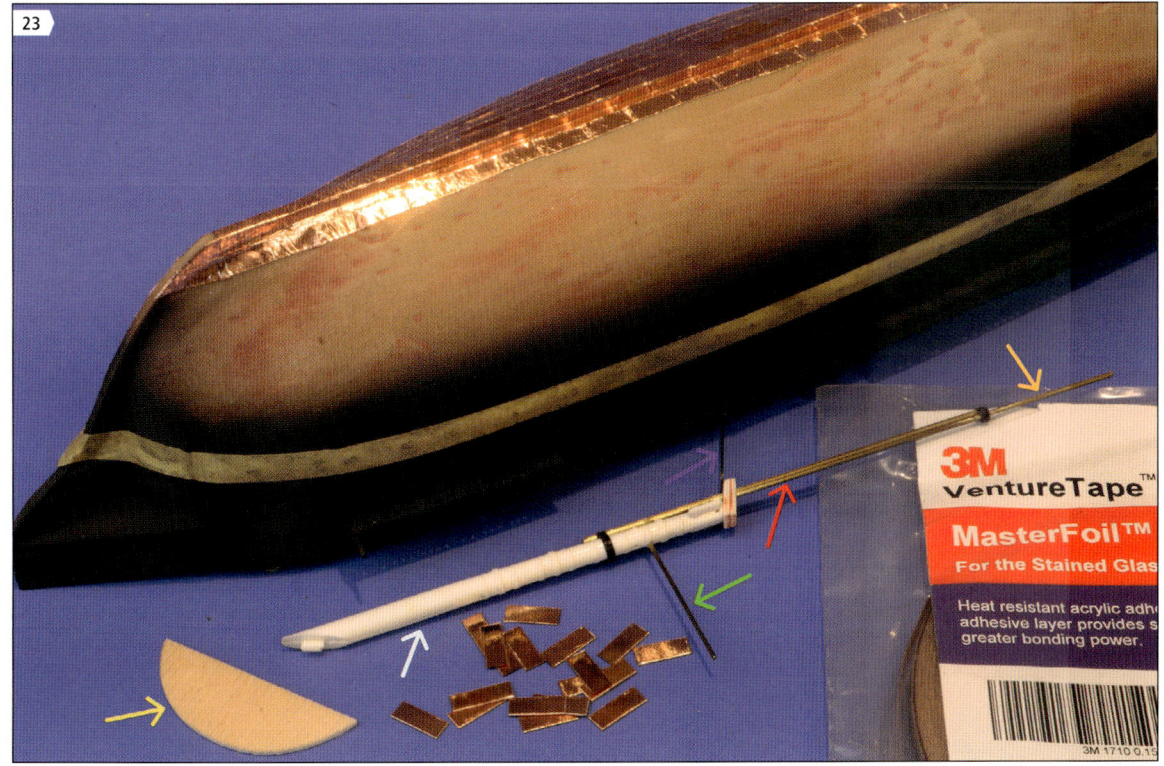

pattern on the surface of the copper. The copper plates are applied to the hull using a thick CA glue. Another popular option is to lay rectangles of self-adhesive copper tape. Copper tape comes in many different widths so it is easy to find a roll to suit the scale of your model. The tape is cut to short lengths using a guillotine-type chopper tool or with a craft knife and ruler. Before cutting the strip many modellers use a ponce wheel to add nail head detail to the tape. Ponce wheels are found at millinery stores for transferring clothing patterns to cloth for cutting. In the model trade, ponce wheels are often called 'riveting tools' used to add rivet details to model aircraft surfaces. Application of copper plate to a hull is easy – just peel off the backing paper and stick. Note that the hull should be sanded smooth before applying the copper plates or else any blemishes or existing detail will show right through the copper.

It is a fun challenge to lay rows of copper because the rows (*strakes*) will not always lie parallel to one another as you work up the hull, but will splay out caused by the curvature of the hull. This will leave gaps between the rows of copper that will have to be filled in with copper pieces cut to shape. This is exactly what a real shipwright had to do, and by laying copper you will gain a real depth of understanding of the complex shape of the ship's hull. Be aware that there are specific rules to laying copper. For example, each copper plate slightly overlaps the previous one on two edges. Depending on which edges overlap will determine where you start your coppering. Does the coppering of the ship start at the keel, laying plates from stern to bow, or does it start at the waterline going from bow to stern working downwards towards the keel? Merchant and naval ships all have different rules and a little research will reveal where to start. Once you have coppered a ship's bottom like a real shipwright, you will intuitively know what to do on the next hull and the job becomes quite enjoyable if not therapeutic. If copper tape is unavailable, a very effective option used by famous miniature ship modellers is to paint a sheet of thin paper or tissue paper for a miniature model a couple of shades of dull copper. The plates are cut from the paper, mixed up, and then glued to the hull in the correct pattern. Once the hull is coppered the whole bottom is given a wash of brown-black paint to kill the whiteness of the cut paper edges. If your model has copper plate detail moulded to the bottom you could sand it all off completely and apply copper plates, or keep it and paint and patina it so that it looks like copper and not copper paint. How this is accomplished is demonstrated in the next chapter.

The Challenge of Shrouds and Ratlines

The one thing that injection moulded plastic cannot replicate well is the miles of rope rigging carried by a ship: in particular, the *shrouds* and *ratlines* that feature prominently on any scale model. The *shrouds* are heavy cables whose job it is to brace each of the masts from side to side movement and the *ratlines* are thin horizontal ropes laced to the shrouds to form ladder rungs for sailors to climb up the masts. These ropes are part of a ship's *standing rigging*, which is essentially part of the ship's basic structure and are tarred to protect the ropes from the weather. On real ships, each shroud is typically set up in pairs. The shroud goes up one side of the mast, around the top of the mast and then back down on the *same* side of the ship. The loop around the mast top is closed with a rope wrapped around the pair of shrouds. Each end of the shroud is *seized* (fixed or tied) to a round, flat block of wood with three holes (*deadeye*) that is referred to as the *upper* deadeye. Each pair of upper deadeyes has a matching *lower* deadeye pair that is attached to the hull by way of an *iron strop* (a band of metal that is wrapped around the lower deadeye). The strop is attached to one of a set of *chain plates* (iron bands bolted to the hull) that are fed through a slot cut into a platform (a *channel* or *stool*) that sets the lower deadeye off the hull side.

The upper and lower deadeyes are laced together through the three holes in each using a lanyard, and by tightening or loosening the lanyards the tension of the shroud is easily adjusted. This set up assembly is repeated for each mast. For the topmasts, the lower deadeyes are attached to the fighting top. Their deadeye strops are fitted through holes in the top and instead of chain plates, their strops are seized directly on to the shrouds. The topgallant shrouds do not require deadeyes. Instead these shrouds are seized to the top of the topgallant mast and their lower ends pass through a trestletree that serves the same function as a hull channel. With the shrouds set up the ratlines made of thin horizontal ropes are laced to the shrouds.

Doesn't this all sound very complicated? It is no wonder model companies have tried different ways to simplify all these ropes with a single plastic moulding that you can simply glue to the hull and masts. However, because it has been virtually impossible to mould scale diameters of rope, these parts are heavy and thick, with shrouds and ratlines moulded to the same thickness. Moreover, there is no way a plastic moulding has been able to capture the delicacy of rope and its *catenary* (sag) caused by gravity and the wind. In an attempt to overcome this, a few manufacturers have included pre-woven thread ratlines and shrouds that have been covered in a thin coating of rubber. Despite their fineness, the shrouds and ratlines are provided with the same thickness of thread and wholly useless. The latest kits now provide a notched loom over which you wind thread of different thicknesses. Once the shrouds and ratlines have been wound it is coated in thinned PVA glue to stick them together. They are cut loose from the loom and glued to the back of a deadeye, then to the top of the mast. The trick to using a loom is to get the shrouds to lie at the correct angle in relation to the masts. The kit instructions will tell you which notches of the loom to wind the thread to get the proper angle and spacing between the shroud and ratline

THE MODEL MAKER'S CHALLENGE

24
The deadeye, shroud and chainplate detail on Revell's 1/96 plastic kit of *Constitution*. The kit parts were replaced with wooden deadeyes and rigged prototypically. When rigging is finished, be sure to check to ensure all of the loose thread ends are trimmed away. The builder Victor Yancovitch missed (red arrow) one that was later trimmed away after spotting it in this photo (used with permission).

25
This illustration shows the upper and lower deadeyes and the order in which the lanyards are laced through them. (*Used with permission from Vanguard Models*).

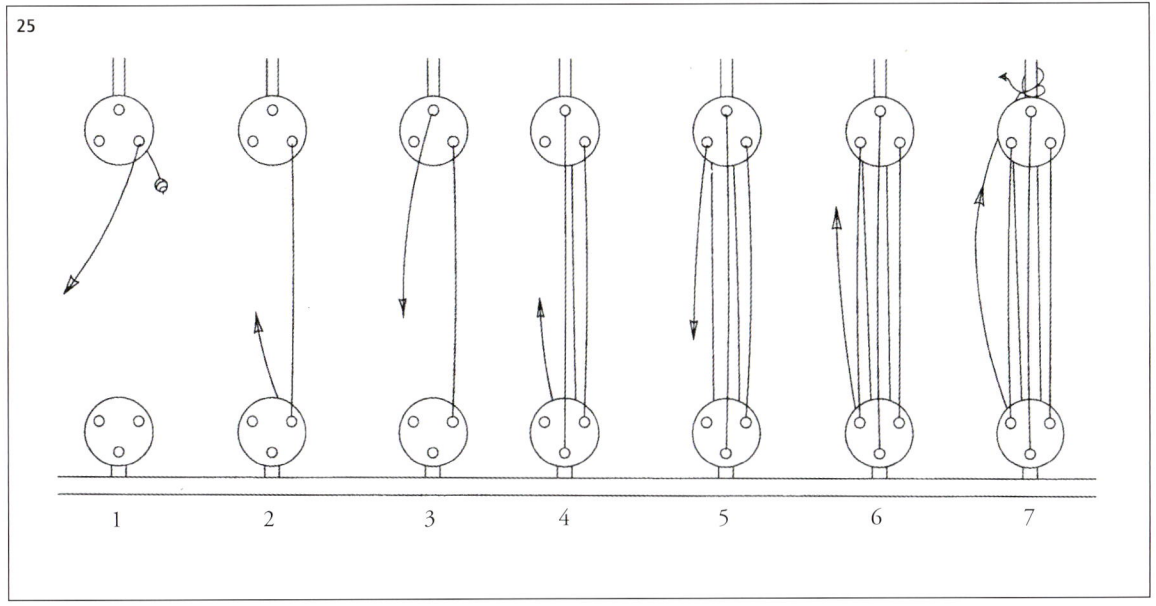

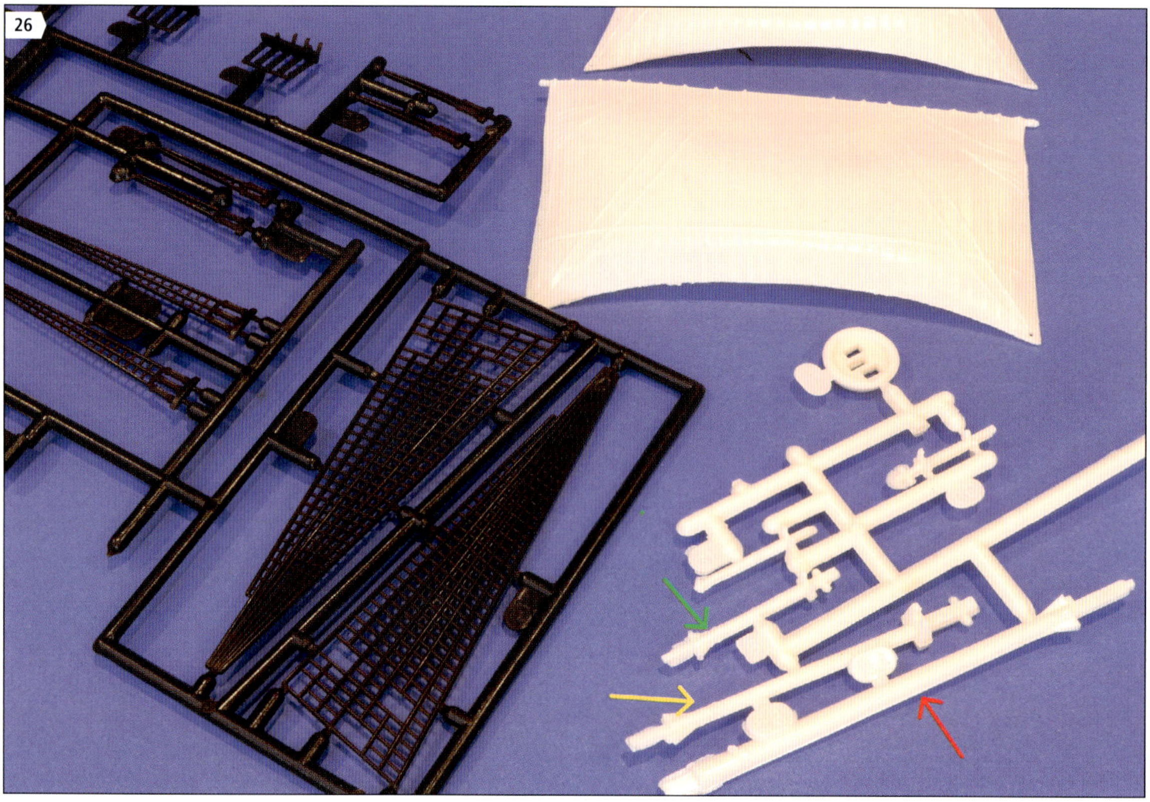

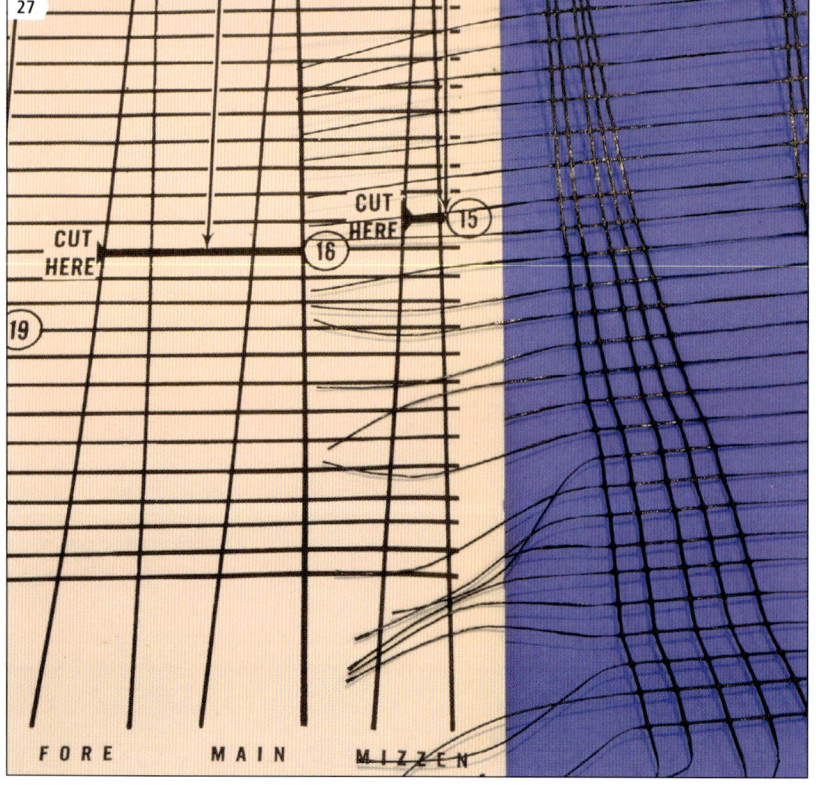

26

These parts are from Aurora's *Sovereign of the Seas* kit. The ratlines and shrouds are moulded as a single heavy part. Although easy to assemble, they are unrealistic. Also shown are parts for a mast – the lower mast (red arrow) is robust enough to withstand the strain of rigging, but the topmast (yellow arrow) and topgallant mast (green arrow) are too thick and unsightly. Replacements from tapered wood dowel will give a delicate look but at the same time imparting strength under strain of rigging. The sails are heavily moulded items best replaced with ones made from paper using the kit parts as a guide to shape.

27

Thread ratlines and shrouds coated in rubber. The shrouds are far too thin.

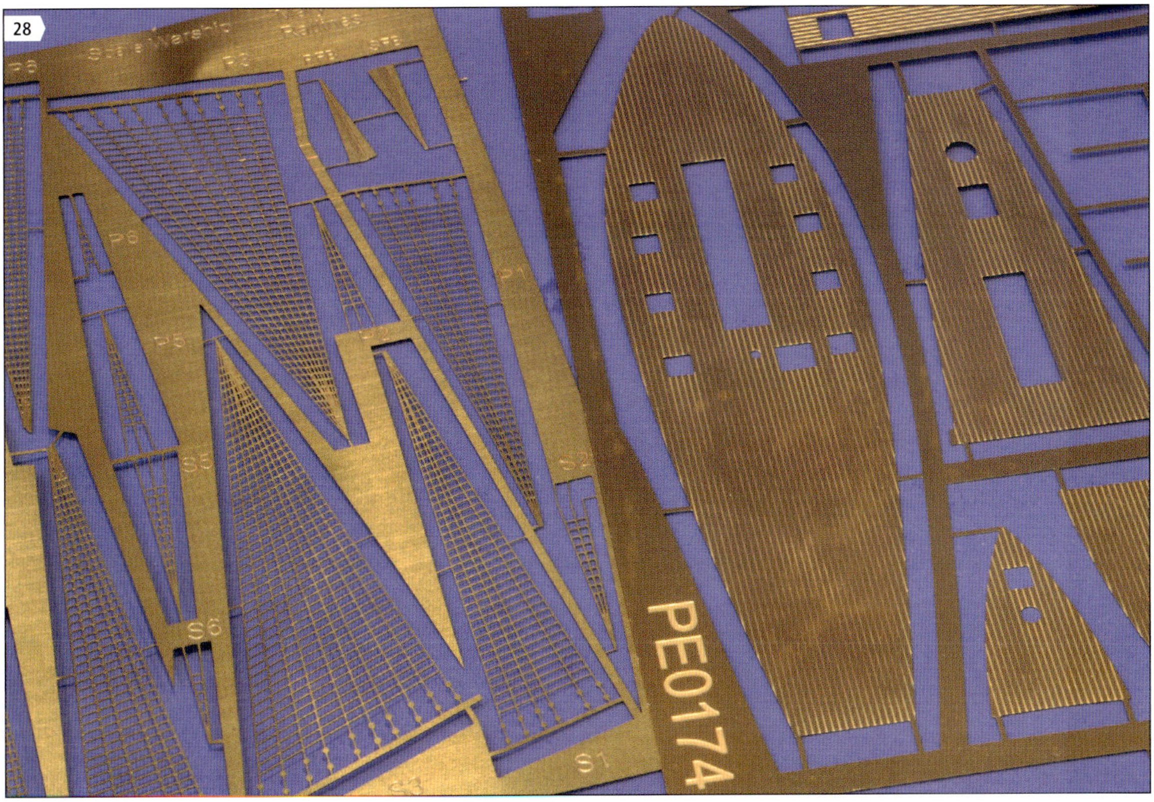

THE MODEL MAKER'S CHALLENGE

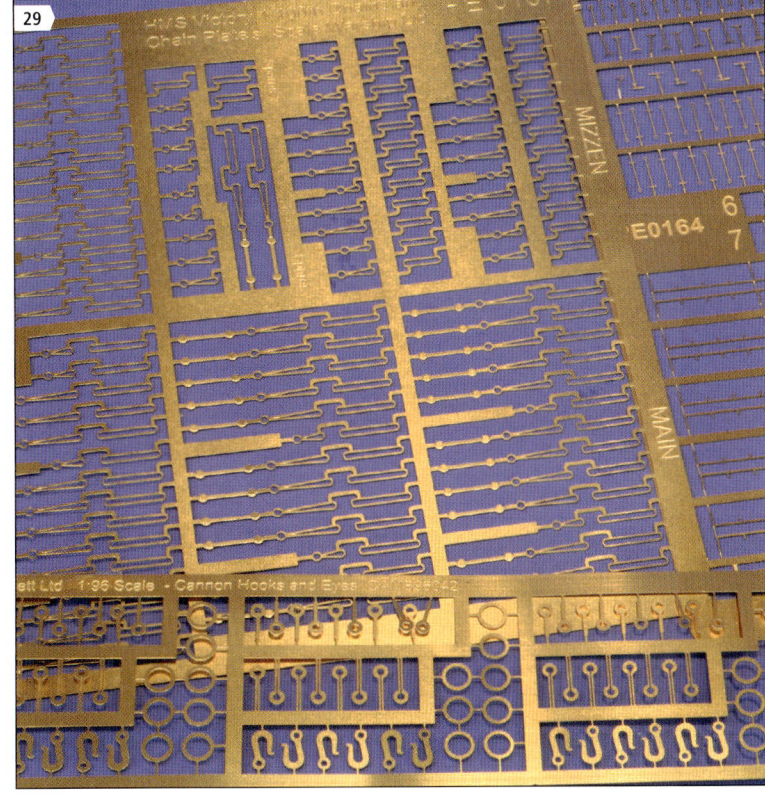

28
ScaleWarship's dedicated PE set for Airfix's *Mary Rose*. The shrouds and ratlines are a single unit with the upper and lower deadeyes. The items are to scale, and the heaviness of the shrouds stands out against the delicacy of the ratlines, yielding a realistic and authentic finish. It is hard to imagine trying to rig this tiny model with thread. To the right are brass decks to replace the kit items. The sets also include other hull and riging detailing parts.

29
Three ScaleWarship PE sets. Top left is a dedicated set of chain plates for Heller's 1/100 HMS *Victory*. On the right is a generic set of mast and yard fittings to 1/150, but can be used in a multitude of scales. At the bottom is a generic set of hooks, rings, and eyebolts designed for rigging cannon, but will find use all over a ship.

threads, but if you glue your masts into the hull at the wrong angle (often caused by some slack in the hole in the deck and the *step* at the bottom of the mast which it plugs into), the windings will not fit.

On small-scale kits photo-etched brass components (PE) are a handy solution to thickly moulded parts. Most of the sets currently available are designed to fit a specific kit, but the parts can be adapted for use on other models. Care must be taken to get the masts fixed into the hull at the correct angle (also called *rake*). The stiffness of the PE parts makes it useful to help set the mast's rake, unlike a thread weaving that is floppy by its very nature. The advantage of these sets is that the shrouds and ratlines are to the correct thickness, but the weakness is that they are flat. This is not a problem for small-scale models but is noticeable on larger scale models. In this case, it is best to use thread and rig them like a real ship. Before using any thread or line for rigging, it is important to remove any hairiness. The most common way is to run the thread through a block of natural beeswax, followed by running the waxed thread through your thumb and forefinger a few more times after that. The heat and pressure of your fingers will soften the beeswax and lay down the fibres, yielding a clean looking line. Some have advocated running the thread quickly through a candle flame to burn off the hairy threads. In my experience this method does not work well, and with all the flammable products on the workbench like solvents, glues, and paints, is a fire waiting to happen. Beeswax can be found at sewing and fabric shops, as well as craft shops where it is the primary ingredient for candle makers.

The standing rigging also includes a number of *stays* rigged forward of each mast (*forestays*) and aft of each mast (*backstays*) that supports each mast from falling backwards or forwards. On real ships, the stays are looped around the mast and tied off, with the end of the stay attached to a block and tackle fixed to the deck or bitt. The stay is tensioned by pulling on the block and tackle. Stays are straightforward to rig on a model, but great care must be taken when tensioning them because plastic masts are easily pulled out of shape. Often after one stay is rigged when the next one is put on the first one sags because too much tension has been put on the second one. Keep an eye out for this as you rig or else you will be cutting all the lines off and have to start rigging

30

The effectiveness of PE shrouds and ratlines (green arrows) are shown here on Airfix's 1/180 HMS *Victory*. ScaleWarship has created several sets for this kit that includes hard-to-replicate items like hammock nettings (red arrows), gratings, deck houses, and open stern gallery windows. (*Photo used with permission from ScaleWarships*)

THE MODEL MAKER'S CHALLENGE

31
Two main mast forestays are shown rigged with thread with 3D printed blocks and hearts (red arrows). Also shown in place are parts from ScaleWarship's PE detail sets providing shaped bow netting, gratings and walkways, and hammock nettings. The effectiveness of mixing PE rigging with rigging made from thread is clear. The trick is to pick the right place to use one or the other. (*Photo used with permission from ScaleWarships*)

the stays all over again. Tensioning is particularly a problem when fitting stays to the topgallant masts. These masts are very slender and even if they are remade from wood or brass rod, they are prone to distortion. An alternative is to use elastic rigging thread used for model biplane rigging or wires strung between power poles on a model railway layout. These elastic lines stay taut without imparting any undue tension. These threads come in several different sizes and colours and are effective in the smaller scale models.

The Challenge of Running Rigging and Sails

Running rigging refers to any rigging that is used to work the ship. These lines are untarred because they are being constantly pulled through blocks. For example, there are series of lines called *halliards* (or *halyards*) and *lifts* that are used to move the yards up and down the mast. These lines are under tension because they hold the weight of the yard in position at the top of each mast. *Braces* are lines that attach to the ends of the yards and run down the mast through block and tackle to belay to a pin rail or cleat on the inner hull sides. These lines are worked to pivot each yard to the best position to catch the wind. The modelling challenge is not just to avoid pulling the yards out of shape, but also to note that being thinner lines they are under greater influence of gravity; the wind acts on them causing them to bend into a natural sag or catenary. Unlike standing rigging in which all lines are taut, with running rigging they can be taut or slack, and the challenge is knowing when to model the sag.

Finally, there is the matter of sails. The modeller can build the ship without any sails in what is called 'harbour rig'. Modelling a ship in harbour rig also means that much of the running rigging is removed because there are no sails to work. If a model is to be built with sails, the first task is to pick which sails. Again, this is where some knowledge of seamanship is helpful because different sails are set for different weather conditions and what the master of the vessel is trying to achieve. Is the ship just cruising under topsails on a fine day in the Mediterranean or is the ship bearing down at best possible speed with the wind at his back towards the enemy? Even a ship in harbour rig may have some sails *bent* (attached) to the yards but are furled, so any lines controlling them will have to be rigged. Having the

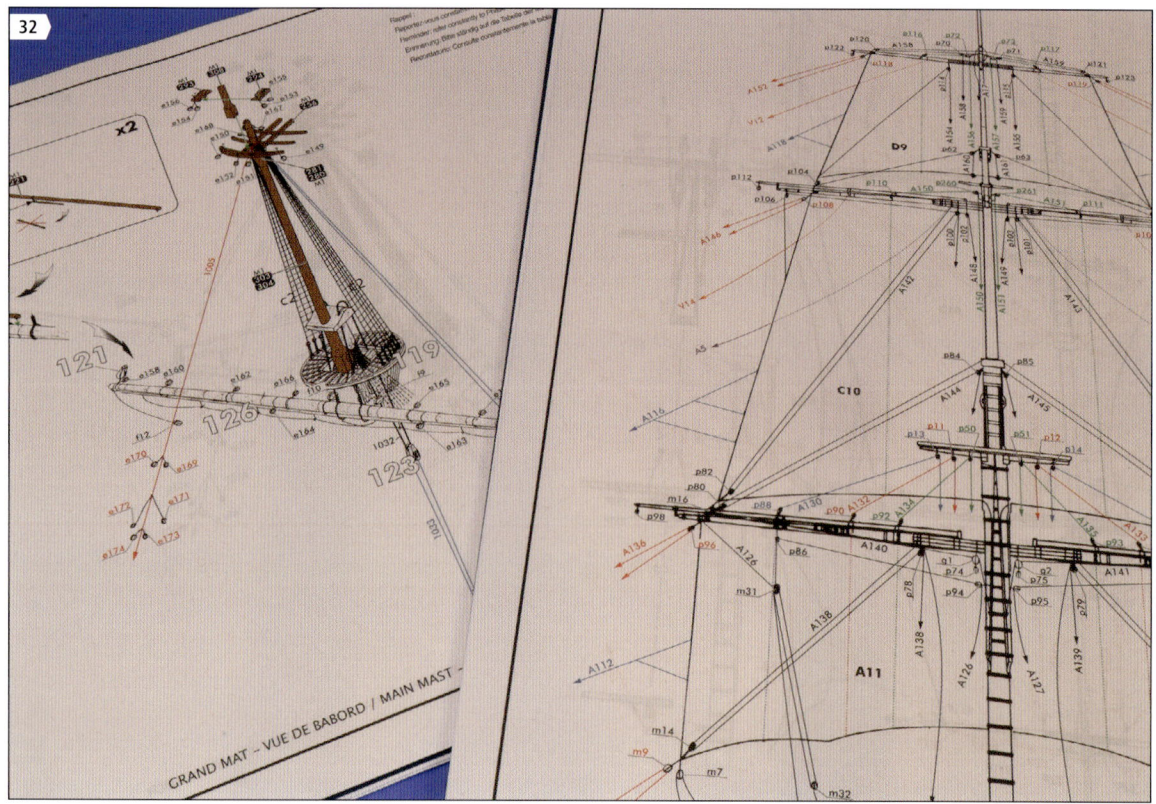

32

Heller's new instructions for 1/100 *Le Soleil Royal* and *Victory* contain updated rigging diagrams that break down the standing and running rigging into digestible steps. On the left is a diagram of the fitting of the upper shrouds and ratlines, the trestletree, and the location of blocks for the running rigging on *Le Soleil Royal*. To the right is a page from the *Victory* booklet showing the run of some running rigging and the blocks they reeve through. Each line is numbered so they are belayed to the correct location on a bitt or pinrail. The rigging shown in these instructions are generally suitable for any model of the same period.

correct sail set on your model is very important, especially in seascape dioramas because they convey the state of the seas and wind, tell the viewer what the ship is doing, and most of all, they bring the model to life.

Kit manufacturers provide sails that would look good on the model, but are not necessarily realistic. The problem is akin to models of jet fighter aircraft. All possible missiles, gun pods, bombs are provided but there are some combinations of this ordinance that is never carried on particular operations. The same consideration goes for sails. When would all possible sail be set? Would cruising under just the topsails be sufficient with the rest furled? How much should the sails billow? The plastic kit sails are often advertised as 'billowing' sails, and often the moulded billowing effect is quite extreme. This would mean that the wind is strong and fewer sails need to be set lest the masts and yards come under too much pressure and snap. The question you must ask is the same question a ship's master would ask. What is the state of the sea and wind? How are sails coping with that wind? What sails should I take in or let out? This will help you choose the correct set of sails to fit to your model.

With kit plastic sails you have fewer choices. The set of the sails and the state of the wind is moulded into them. The choice of which to fit is very much determined by what is provided in the box so you may be leaving some of the kit sails off, but may use most of them. It all depends on the answers to the questions you are asking yourself. Moreover, care must be taken with the stiff plastic sails because they often do not hang naturally. Yes, they have moulded sag, but without care, they look like they were added to the model, and not *part* of the model. This is caused by any slight, almost minute angle that would alert the viewer that the sail is defying the laws of gravity. Despite having a moulded sag, when they are attached to your model, the bottom of the sail may just jut forward or backwards a little in seeming defiance of gravity. Similarly, the degree of billow moulded into the sail suggests that the face of the sail should protrude forward because they are full of air. If you attach them to the yard at the wrong angle, it will look like the sail got out of control or the rigging used to control the shape has not been handled is a proper seaman-like fashion. Real sails are huge sewn sheets of canvas that weigh tons and hang a particular way off the yards. The state of the winds, if the sails are wet, and how the lines controlling them are hauled in or let out will have an impact on its shape. Plastic sails limit your choice and once attached to the model are difficult to shape.

Many modellers have tried to make sails out of cloth to better capture their look. The challenge is scale. Fabric has a distinct weave and the trick is to find one with a very fine weave. The finest weave fabrics are found on high quality cotton dress shirts. Fabric is also stiff and will not hang naturally because the fabric has been treated with a sizing to keep it wrinkle free for the wearer. Any sizing can be removed with several washes in hot water, and the physical action of washing helps wears the fabric down so it is easier to shape and makes it softer. Fabric sails will have to be dyed to kill the bright whiteness of the fabric they are made from. This is a messy job, and after all this, they still have to be cut and sewn into sails. There are many modellers who can work a sewing machine but the stitching also has to be in scale, otherwise the sail will look coarse and unrealistic. Many home sewing machines, unless they are very high end with multiple functions, cannot produce the fine stitches modellers require. This is why fabric sails are most often found on large-scale models because the size of the ship better matches the weave of the cloth and stitching. For smaller scale models, the alternative is to make new sails out of paper. Paper has virtually no grain, is easily painted and coloured, and after the sails are attached to the yard, paper is easily shaped to hang naturally. A new development for smaller scale kits is CNC cutting and sewing of cloth sails. A computer can cut and stitch the sails to much finer sizes as a better match to a smaller scale kit. The fabric used in these sail sets are dyed a light tan colour making them useable out of the packet.

33

Vacuformed plastic sails can be used with care, but they cannot be shaped after mounting to the model. This set from Revell's *Cutty Sark* (primed tan for clarity) even has moulded *buntlines* (lines used to lift a portion of the sail – red arrows) and clew lines (lines used to manage the edges of a sail – green arrow) that do not look realistic at all and its flatness does not resemble the shape of the thread used for the rest of the rigging, no matter how artistically painted. The vacuform sails are ideal for use as a mould for new sails made from lightweight paper. Paper is soaked in a mixture of white glue and water and laid over top of the vacuformed sails. Once dry, they can be popped off the mould, cut to shape, and coloured. Moulded paper sails remain flexible and can be shaped to hang correctly after mounting to the ship. Proper buntline and clew lines from thread can be rigged as well.

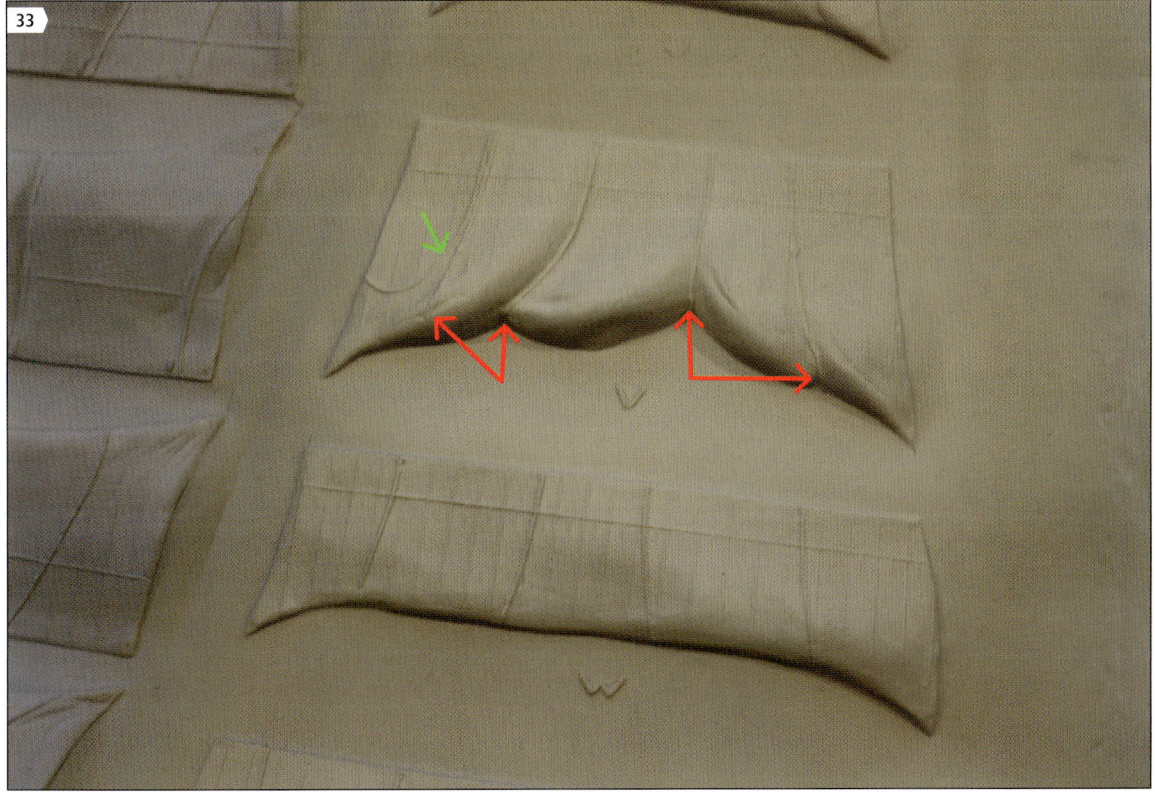

Most kits will advise you to paint the vacuformed sails a solid white or off-white colour. Some kit instructions will say to weather them with a wash of tan paint. However, sails were not made of a single, huge piece of canvas. Rather, they were sewn together with strips of fabric and each strip was a slightly different colour and weathered differently. In this case, some thought will have to be given to recreating this patchwork finish for a really authentic look. Some kits even mould rigging lines onto the face of the sails. This looks terrible because the moulded rigging line will look nothing like the other rigging lines made from thread. Worse yet, because the sails are vacuformed the moulded rigging line is convex on the face of the sail but a concave groove in the back side!

Pennants and flags are very important aspects of a sailing ship and care must be taken to ensure your ship has the correct ones hoisted. Ships of the sailing era communicated entirely by flags, not wireless, so having the correct flags flying is important. Many kits include the wrong ensigns. It is interesting that kit makers all include a modern White Ensign for any British ship even if she never flew one. More subtly, the Grand Union often provided is a modern one that does not take into account when the Kingdoms were united. If it looks British it must be British it seems. American flags are also poorly served. Many kits provide a flag with all 50 stars, ignoring the gradual accession of states. Sometimes, an archaic version of the flag (*eg*, the flag reputedly designed by Betsy Ross) is included because it looks American. After finding the correct flag, the model making challenge is how to make them look realistic as opposed to a piece of printed paper glued to a mast. They must be of a scale thickness and appear to flutter in the wind as a real flag does. If using cloth flags, again, the issue of the size of the fabric's weave the design was printed on must be taken into account before deciding to mount it to your model.

These considerations and how to tackle them is the fun part of model making. The scale of the model will determine the types of techniques that can be used. In general, techniques used by miniature model makers Phillip Reed and Donald McNarry (their books are listed in the Resources section) are perfect for small-scale models, while larger models can adopt increasingly prototypical techniques used by wooden ship modellers. However, some modellers will want an out-of-the-box build with just some simple refinements, while others desire greater detail with the addition of scratch-built parts, and/or the inclusion of after-market detailing parts. Keep in mind that whatever way you choose to build your model is perfectly acceptable – and not to be discouraged by the 'detail snob'. What one person thinks is a problem, such as simulated woodgrain finish because it is invariably over scale, is perfectly acceptable to another who believes that a model of a wooden ship should looks like it was built from wood. There is no right or wrong – it's your choice.

34
The back side of vacuformed sails. All the detail moulded onto the sail is now unrealistically recessed.

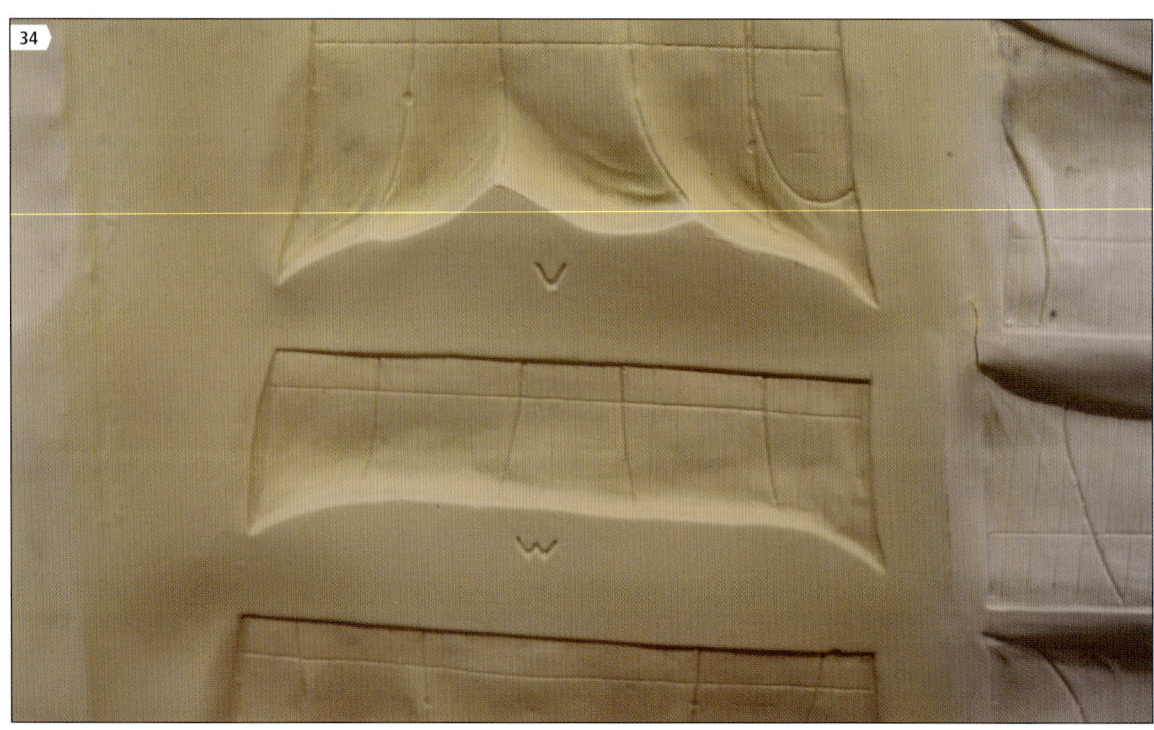

35

Langton Miniatures 1/300 *Victory* is a multimedia kit. The kit sails were etched brass sheet, but were used as a pattern to draw a new suit of sails on copier paper. The seams were drawn on both sides of the paper with pencil and given a wash of light tan paint. Different shades of the wash were painted between randomly selected seams to give the impression the sail was made from individual bolts of fabric. The kit supplies the shrouds, ratline, and deadeye assemblies as etched brass components that work well.

Given the small scale of the model, just representative rigging was added but the purpose of each line was clear. Fore and backstays are from different sizes of black thread. The running rigging was limited to braces and made from tan painted wire. Wire was selected because after mounting a realistic catenary can be worked into them with the back end of a paintbrush handle. Of note are the stunsail booms added to the ends of the yards, and the furled stunsails made from damp tissue paper that was twisted up and painted tan. The St George's Cross on the foremast is hand painted onto fine paper and bent to look like it is fluttering in the wind. Figures are from the Langton range meant for wargaming in this scale. Unfortunately, Langton's *Victory* is no longer available but they have a wide range of other ships from frigates to chebecs in this scale, and plenty of seamen, officers, and marines to populate them.

3: Building Out-of-the-Box
Victory, 1765 Airfix 1/180 Scale

Our first build will be Airfix's venerable 1965 kit that remains available to this day. Although the boxings are emblazoned *HMS Victory 1765*, the ship is definitely not the open stern version as launched, but the actual kit replicates her 1950s restoration to 'near Trafalgar' appearance as a museum ship in Portsmouth. The kit appears to be based on the drawings found in C Nepean Longridge's 1961 book *The Anatomy of Nelson's Ships* drawn by George F Campbell. This is a good thing because at the time these were the best drawings available and remain valuable today. When Airfix started releasing the new range of sailing ships, they proved to be a bestsellers, so in conjunction with the publisher Patrick Stephens, a series of books entitled 'Classic Ships, Their History and How to Model Them' was released. The books were authored by Noel Hackney, and the book on *Victory* was released in 1970. Others included the *Cutty Sark* and *Mayflower*. This book remains helpful today because he identified what is missing from the kit, and the rigging plan for the ship in harbour rig is very complete, down to the correct belaying pins for each line. If you follow this book you will get a very highly detailed model of the ship but it will take considerable skill to achieve all that Hackney did. Indeed, many who have tried to replicate all of his efforts gave up and never finished the model. I am one of them!

In the intervening 55 years, a very large range of aftermarket accessories have become available for the kit, making scratch-building the way Mr Hackney did somewhat redundant. These accessories replace any poor kit parts or offer parts that are missing from the kit. These items include fully detailed wooden decks in oak or beech veneer with all the planking, butts, and joggling laser etched into the surface. There are PE rigging sets for the shrouds and ratlines, and sets with all the miniature real wood deadeyes and blocks to be *reeved* (threaded) with multiple gauges of thread. Being a ship of the line, sets of turned brass guns and laser cut carriages are available, along with full suits of fabric sails and accurate flags. With these sets the kit can be fully outfitted for a superb miniature. Obtaining all of these sets also amounts to a considerable outlay of cash, and for someone new to sailing ships it will be daunting task to use them all to good effect.

With this in mind, the build presented here will use some simple and inexpensive plastic modelling techniques to refine the Airfix kit and produce an authentic model of *Victory* out-of-the-box. In general, the kit mouldings are quite good with a delicately detailed hull with carvings and moulding and

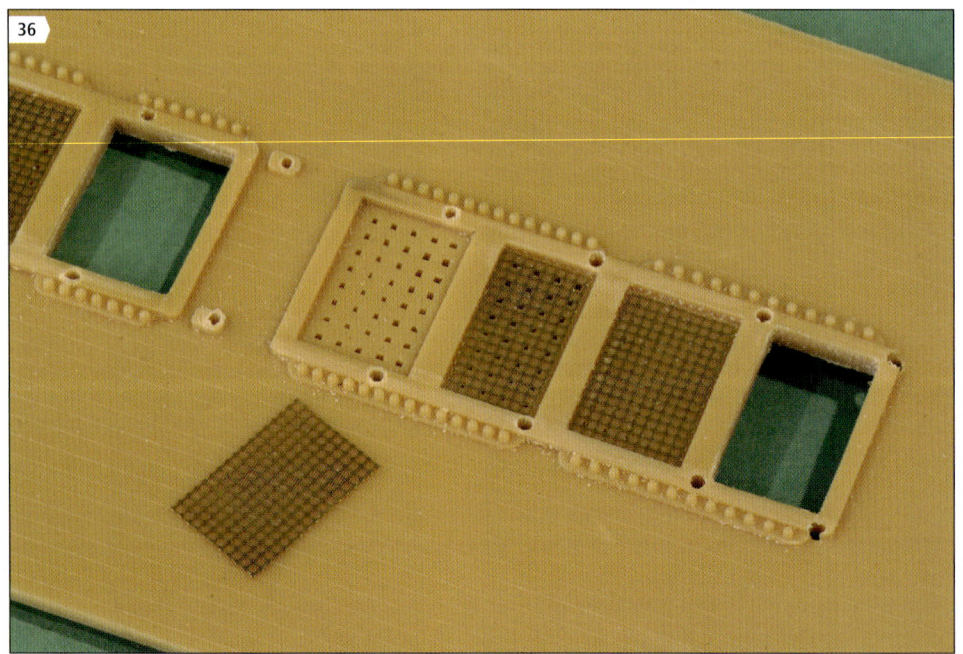

36

The first step is to improve the deck gratings with etched brass mesh left over from another model, but any kind of screening material or fabric netting will work. The key is to use a material that has the right sized holes in it.

BUILDING OUT-OF-THE-BOX

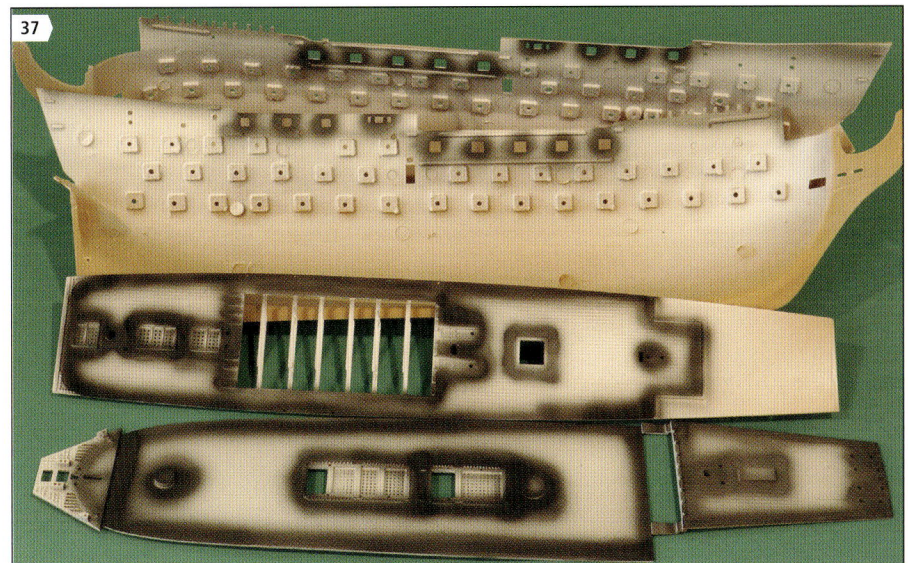

37
The kit parts were cleaned up and primed white. In preparation for painting the interior of the ship, the decks and interior bulwarks are pre-shaded in dark grey paint.

38
Victory's decks are shown here painted and ready for installation between the hull halves. The decks were painted with thin coats of a light grey-brown colour (Tamiya Deck Tan XF-55) by airbrush, allowing a little of the dark grey pre-shading to show through. The deck colour was lightened with white paint and sprayed on the deck centres to create highlights. The interior hull bulwarks were sprayed with yellow ochre (Admiralty Paints AP9115W) allowing some pre-shading to show through. The paintwork was given a thin wash of artist's oil black and burnt umber oil paints dissolved in odourless thinner. The wash blends the shades and brings out any moulded detail.

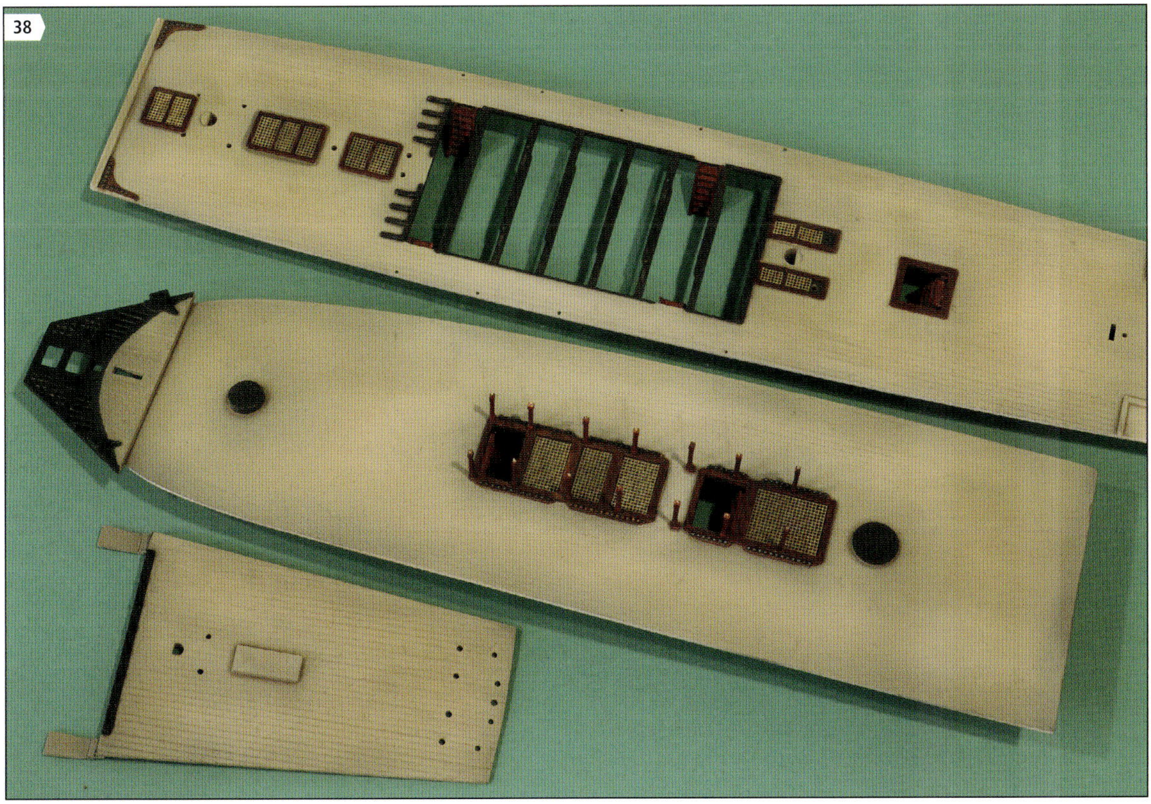

copper plating detail rendered in the correct pattern. The hull planking detail is heavy, but the main *wales* (strakes of wooden planking that is thicker than the hull planking to give the hull strength) carry delicate 'anchor stock' planking detail like the real ship. The side and stern galleries are separate parts that feature detail proud enough to be easily picked out by drybrushing and tidied up with a fine paint brush. Fine raised lines represent deck planking, but the hatch coamings are detailed with over-scale gratings and are prime candidates for refinement. The boat skids in the ship's waist are thin planks instead of substantial square section beams, and these too can be easily cut away and replaced.

The lower masts are robust enough to take the strain of some rigging, but the topgallants and topmasts may need replacing with wood depending on the method of rigging you choose. The yards are very fine, and topgallant and spritsail yards should definitely be remade with brass rod or shaped bamboo skewers. The bowsprit is stout but the jibboom and flying jibboom should be replaced with something stronger. The fighting tops are detailed and follow the Campbell plans precisely. No studdingsail booms are included. The deadeye pairs and lanyards are moulded as a single unit to the channels. A loom is provided for the shrouds and ratlines that are glued to the back of the upper deadeyes. The rigging diagram provided in the instructions is very basic and probably the greatest weakness of the kit.

The biggest change to the kit will be to recreate the standing rigging emulating methods used by real ship's riggers, but modified for model making purposes. *Victory* is well covered by hundreds of books, but the best ones for a model are John McKay's *Victory* in the Anatomy of the Ship series, this author's *Shipcraft 29: Victory, 100 Gun Ship*, the aforementioned Longridge's book, and Lennarth Petersen's *Rigging Period Ship Models*. The kit will be built to reflect her appearance in Portsmouth prior to 2016 and carry a simplified harbour rig.

39

The hull is shown assembled per kit instructions and the kit's beakhead deck, roundhouses, and gratings are painted. At the stern the quarterdeck screens were painted in a dark varnished wood finish (Vallejo #70828 woodgrain). Historians at the HMS *Victory* Museum now think that these screens were not dark finished wood but were painted yellow ochre. During the ship's 2016 restoration they were painted as such.

The join between the quarterdeck and the hull was covered with a strip of thin styrene painted black (red arrow) to represents the waterway that was not included in the kit. In the ship's waist, the kit omits a cap rail along the hull sides around the open area (blue arrow). This was made from a strip of .75mm plastic and whose width overlaps the hull side and deck slightly. The edges to the cap rail were rounded off, pre-painted black before being cut to length and cemented in place.

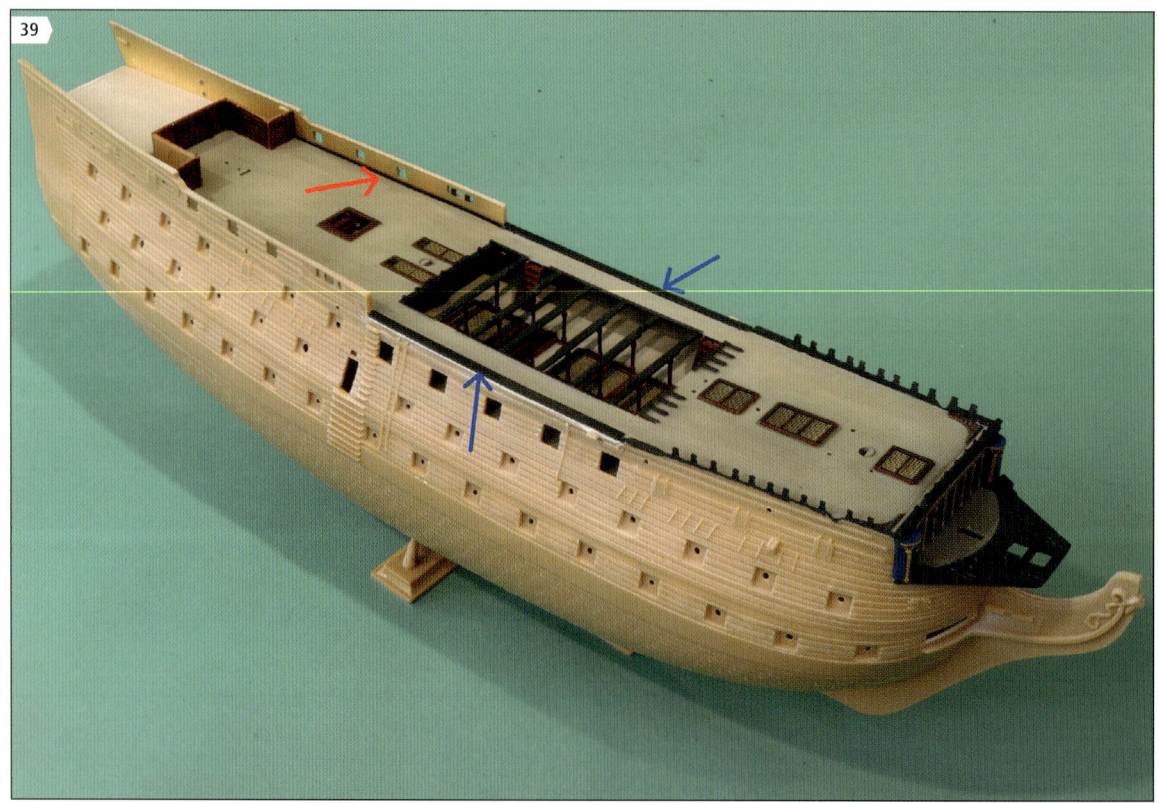

BUILDING OUT-OF-THE-BOX

40

A cap rail was also added along the quarterdeck sides (white arrow), and the cap rail added to the waist can be clearly seen (white arrow). In preparation for rigging the model more realistically, the kit's deadeyes and chain plates moulded to the channel parts (blue arrow) were cut away and sanded smooth. Holes were next drilled in the location of the lower deadeyes. The hull carries some moulded-on chain plate detail and holes were drilled through the hull at the top of this detail (black arrow). These holes will accept a length of thread used for the shroud that will also stand in for the chain plate.

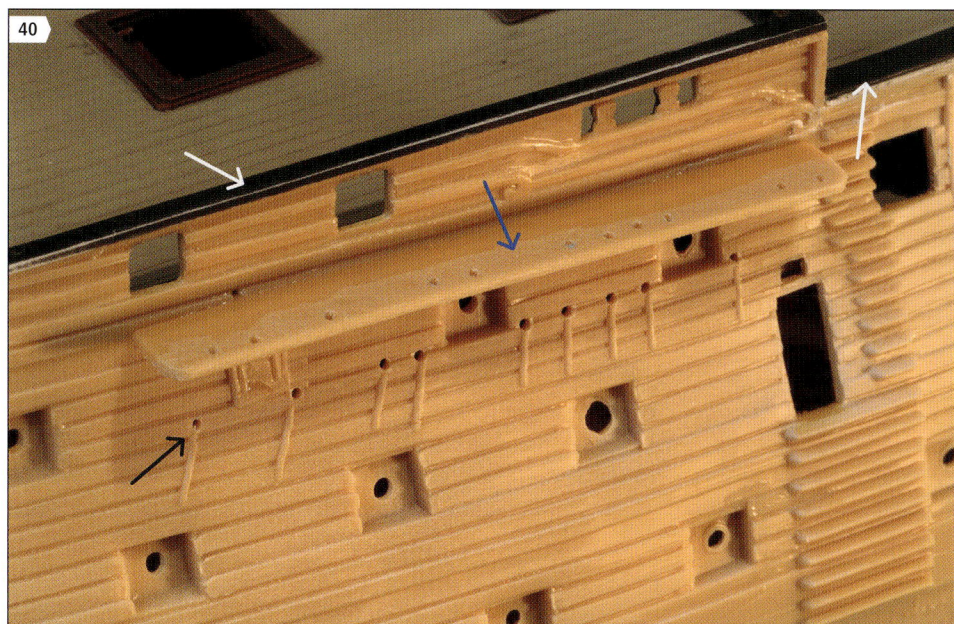

41

The kit omits a small channel, called a *stool* that is present on *Victory* to which the mizzen mast backstays are rigged. The position of the stool was located on McKay's drawings and made from a scrap of styrene. Its location is shown here on the painted hull (red arrow).

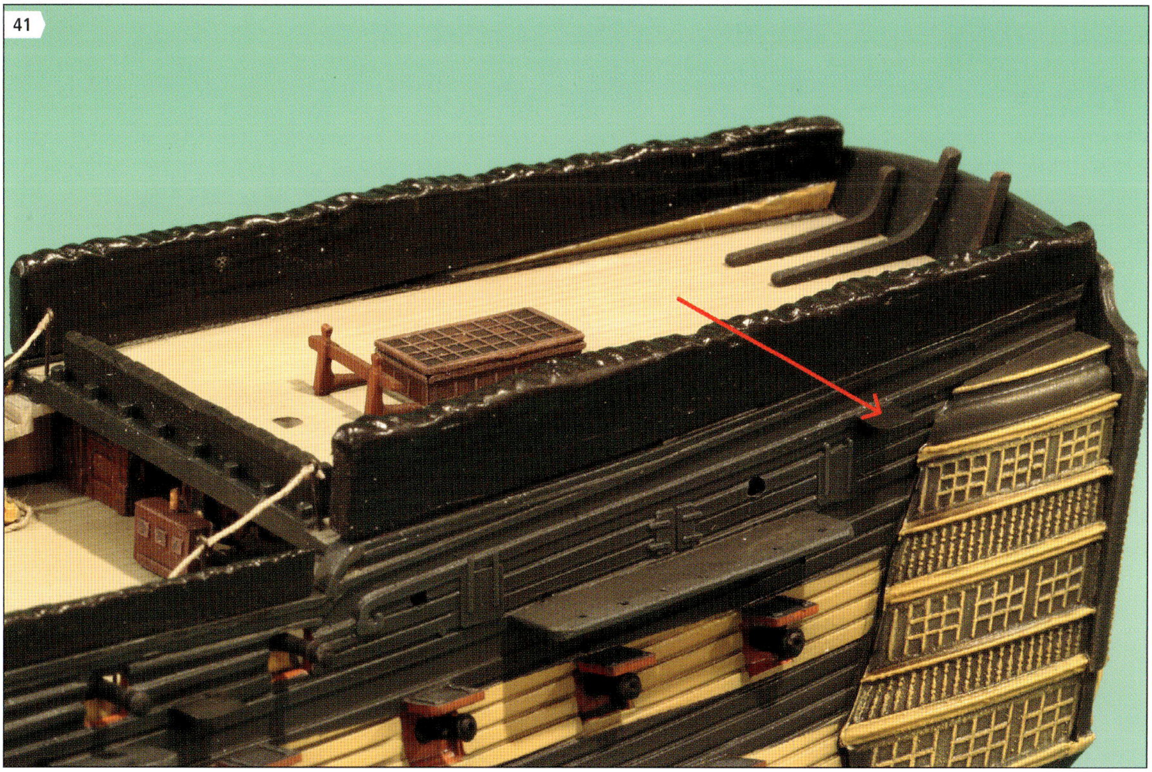

42

The model's exterior was primed white and the bottom painted with Alclad Copper paint (ALC 110). It is bright and shiny so it must be dulled down.

43

A wash of black and burnt umber artist's oil was applied to the copper and allowed to dry hard. A greenish patina was added using verdigris, black, and white artist's oils that were very roughly mixed with a little oil paint medium (*eg*, Newton and Windsor Liquin) that helps speed drying and allows the thick and heavy oil paint to be brushed on smoothly. These colours were randomly daubed onto the copper and a large stiff brush used to brush it out. Additional dots of paint were added – maybe more verdigris, more black, a dab of whitish green here and there, and all is brushed out until the effect you want is achieved. You are aiming for a bit of a patchwork effect with these colours as opposed to a uniformly homogenous finish. Matt varnish protects the paintwork.

44

As with the decks and inner bulwarks, the outer hull was pre-shaded with dark grey and the ochre and black stripes were masked off and painted. For simplicity, the gun ports are painted red ochre but the recess could be painted black and the sides red ochre to better represent how the gun ports would actually look with the port lids raised. To finish, the entire hull received several thin washes of black and burnt umber oil paint and when dry any excess wash is removed with a cotton bud moistened with white spirit. A coat of satin clear varnish seals in the paintwork and deepens the effect of the wash. The hull is looking like *Victory* now.

45

The stern and quarter galleries carry a lot of finely moulded decorative detail that is easily painted by drybushing with yellow ochre. The drybrushing will define the raised detail and you can then use a fine brush to fill in any missed edges. A pin wash of black and burnt umber artist's oil paint was applied around the details to sharpen them even further and cover up any stray drybrushing marks.

46

On ships in service the crew's rolled hammocks were placed in the nettings to air out, and in uncertain weathers were covered by tarpaulins or oilskins. The shapes of the hammock nettings were cut from wood. A sausage of two-part epoxy putty (*eg*, Green Stuff or Epoxie Sculpt) was placed along the top edge with lumps representing the hammock ends sculpted in. On the smaller nettings a strip of wood was cut to the height of the nettings plus the hammock tops. Lumps were filed into the wood to represent the rolled hammock ends. The parts were given an overall coat of artist's acrylic gesso to recreate the texture of a tarpaulin covering. When dry they were painted black and given a semi-gloss finish to convey an 'oil skin' appearance and glued into place on the hull.

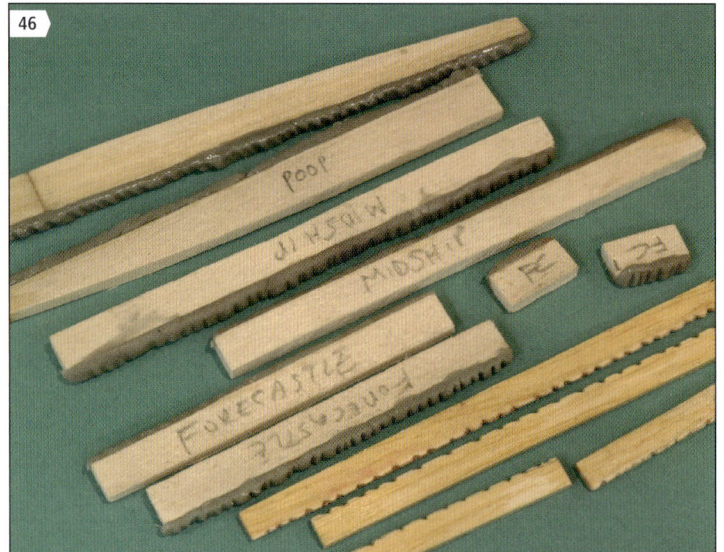

47

The covered hammocks have been fixed to the hull (white arrows). Deck fittings and ship's boats were also assembled, painted and added to the deck. The ship's galley stove chimney was detailed with a disc of styrene to represent its cover (green arrow). All of the visible guns received a breeching rope of thick tan thread (blue arrow). The thread is cut to length and dipped in a solution of thinned PVA glue. The damp thread is laid across the gun's *cascabel* (the knob-like projection at the rear of the gun barrel) with the ends glued to the deck by the front wheels. After all the remaining hull and deck fitting were painted and fitted, a coat of satin varnish finishes off the basic hull assembly.

BUILDING OUT-OF-THE-BOX

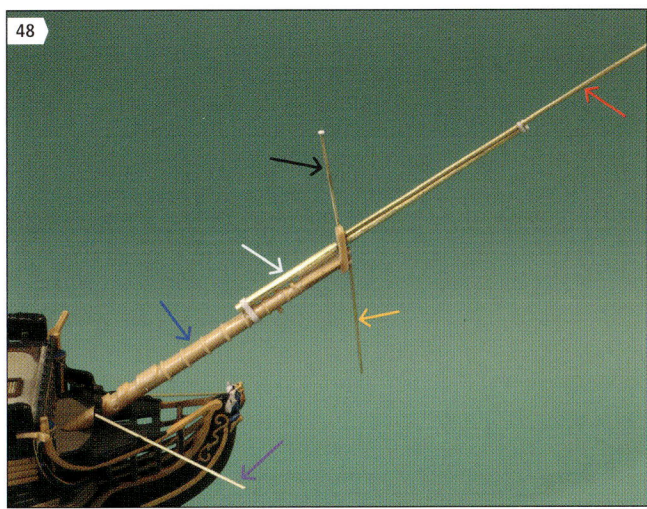

48

48

In general, the kit's lower masts are thick enough to be used out of the box. The bowsprit needs some modification and the moulded jibboom and flying jibboom was cut away. The photo shows the kit's bowsprit (blue arrow) was mated with a new jibboom (white arrow) and flying jibboom (red arrow) made from brass rod. The jack staff (black arrow) and martingale (yellow arrow) are brass rod. The martingale had four notches filed into it to hold the stays when rigged later. These new brass parts will take up the strain of rigging without bending out of shape, and the finesse of them over the clunky kit items makes the effort to make them worthwhile. The *boomkins* (a short spar sticking out from the hull used to attach rigging such as braces) are brass rod replacing the kit's items (purple arrow).

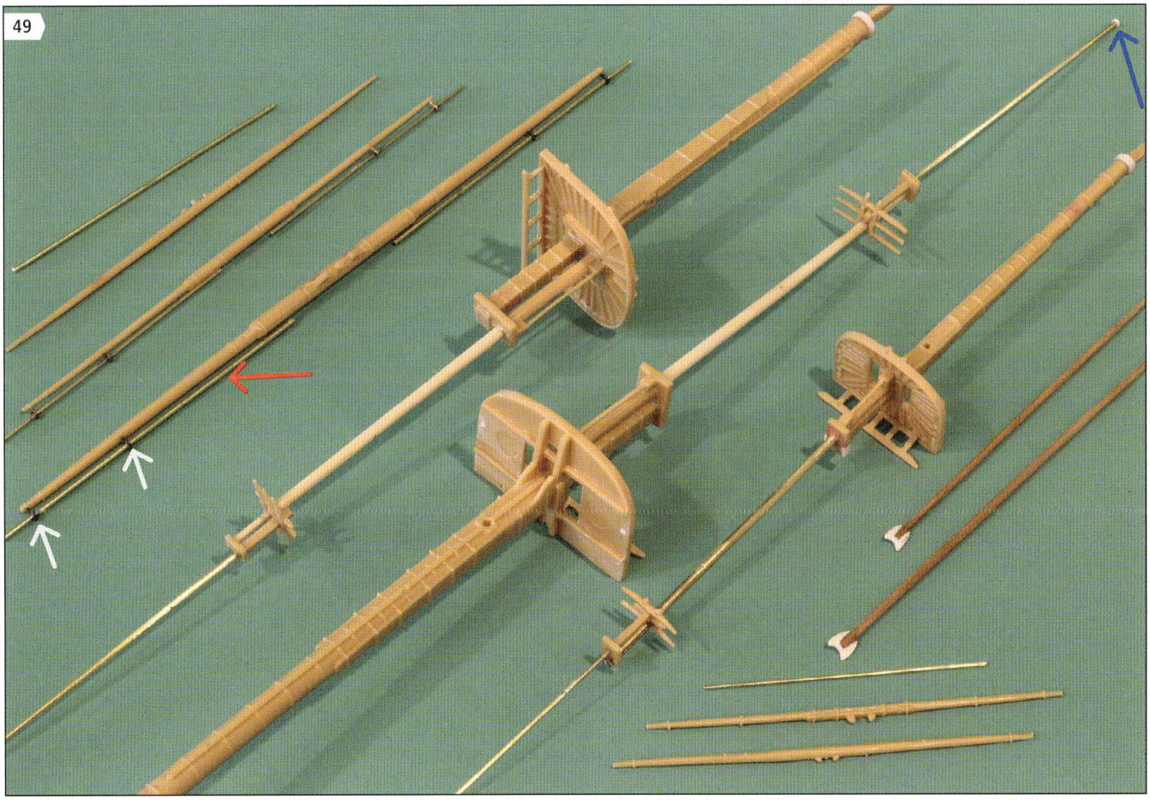

49

49

All the masts received new topmasts shaped from wooden dowel, and topgallant masts from brass rod. Each mast is capped with a *truck* (a wooden ball or disc that carried holes in it through which flag halyards are passed) made out of a disc of punched card (blue arrow). Each mast has a *coat* added to its bottom made from a strip of paper wound round and around the base of each mast. The paper is soaked with thin cyanoacrylate (CA glue) and the top edge sanded to a rounded over shape.

The lower and topmast yards are the kit items with stunsail booms fitted to their ends (red arrow). These booms were fixed to the yards by sliding them through tiny brass eyelets (found at craft stores) set into holes (white arrow) drilled into the yards. The yards and masts were then primed, painted and fixed to the hull.

50

50

The painted masts are *stepped* (fixed to the hull) at the correct angles.

51

The standing rigging begins with the shrouds. The shrouds were fitted by gluing the end of the shroud into a hole in the hull, the other end is fed up through the hole in the channel, up around the mast (where it is seized off), then down the mast and through the next hole in the channel. The end is glued into the next hole in the hull. The thread used for the shroud does double duty as a chain plate and shroud. Dipping the end of the thread in CA to form a needle point will help in feeding the thread through and into holes.

52

The shroud is led up the mast to pass through the mast top and how they are seized to the mast with a loop of thread is shown. The excess is cut away after securing the knot with a touch of glue.

BUILDING OUT-OF-THE-BOX

53
The upper and lower deadeyes are represented by discs of card glued to the shroud. The face of the disc is given a dollop of PVA glue to give it a more rounded shape and painted black. In this scale, the modelling ruse works well and the shrouds have the proper rake in relation to the mast. The deadeyes could be super detailed by gluing three short lengths of fine wire in between the upper and lower deadeyes to represent the lanyards.

54
The same method was used for the topmast shrouds.

55
Let the tedium begin! The ratlines are individually cut lengths of fine thread glued across the shrouds and trimmed off when dry. Try to get them straight and not let them sag as they do on my model (red arrow). A piece of white card placed behind the shrouds really helps you see these tiny threads.

56
This is an example of what not to do. The ratlines are almost as thick as the shrouds and this looks heavy and unrealistic. The ratlines were not pulled taut enough when fixing to the shrouds so they lie unevenly across the shrouds, and several of them defy gravity by bowing upwards.

SAILING SHIPS FROM PLASTIC KITS

57

58

57

Fore stays (red arrows) and backstays (white arrows) were rigged, finishing off the standing rigging. Different weights of threads were used to represent the different stays. On the upper stays and martingale under the bowsprit (black arrow) are black coloured elastic thread (*eg*, EZ Line) was used because this material stays taut without putting any tension on the slender masts. References were used to identify the most prominent stays to include to give the impression of full rigging.

58

As the ship in in harbour rig, only a little running rigging needs to be fitted because there are no sails to manage. Ships spending an extended time in harbour often had any unnecessary rigging removed to save the lines from damp and rot. Representative lifts (white arrow) and braces (red arrow) required to support the yards were fitted. Tan coloured elastic thread was used to keep the lines taut, which is easily fixed to the yards with a dab of CA, requiring no knots to be tied.

59

Bowsprit rigging.

60

Stern quarter view of the completed model. The ship's White Ensign was an image found on the internet, resized and printed out on copier paper. Running the paper twice through the printer gives a dense, saturated colour to the flag.

61
The completed model. Even with simplified rigging and minimal detailing to the out-of-the box kit, *Victory*'s grace and power is clear to see. Careful shading and highlighting of the paintwork throughout the build bring the ship to life.

4: Beyond the Box
HMAV Bounty, 1787 Airfix 1/87 Scale

Captain William Bligh's ship needs little introduction as the setting for one of the most (in)famous mutinies in the Royal Navy. The *Bounty* began life as the collier *Bethia* that was taken into naval service and modified to essentially become a floating greenhouse. Bligh's task was to sail *Bounty* to Otaheite (Tahiti) to gather breadfruit plants and carry them to the British West Indies as a cheap food for plantation slaves. The record of conversion of *Bethia* to *Bounty* is quite complete on the changes required to help ensure the survival of the breadfruit plants on the journey, but also changes to her rig and rigging to cope with the orders that directed Bligh to sail around the Cape of Good Hope and treacherous seas where the Atlantic and Pacific meet.

This build of Airfix's *Bounty* will fit several aftermarket parts and carry out more prototypical standing and running rigging, and fit cloth sails. The kit is well moulded with a lot of fine detail and her shape accords well with the Admiralty draughts of the ship. John McKay's *Bounty* and this author's *Shipcraft 30: HMAV Bounty* are two modeller-friendly books that highlight the features unique to the ship. Firstly, the kit's mast tops are moulded solid, but Bligh had them replaced with *gratten tops* (tops whose floors are fenestrated with open gratings). Second, the kit includes hulls for *Bounty*'s original allotment of three boats (a launch, a cutter and a jolly boat) but on the famous voyage only the launch and cutter were carried. Thirdly, the moulded hatch frame sides are solid and must have holes drilled out to allow greater ventilation for the cargo of breadfruit plants. Additional challenges are that the simplified bitts and the windlass cheeks are unrealistically thin or missing and best remade from thick styrene sheet stock. The kit omits a few pairs of deck bitts, the compass binnacle, lanterns, and belaying racks and pins. Some moulded detail is incomplete, such as the copper plating that is not extended over the stem from the rest of the underwater hull. The copper's pattern is represented with fine recessed lines in the correct pattern but is devoid of any overlap and nail head detail.

The final challenge is that there are no contemporary illustrations of the ship which shows her actual finish. No one is sure of how *Bounty* was ever painted, and the artist's renderings on model kit box tops all show wildly differing colour schemes. In most illustrations she is shown with a payed natural wood finish with black wales topped with a 'yellow racing' stripe, and white underwater hull and masts. One thing for sure is that many kit manufacturers have based their kits and colour details on sailing replicas created for motion picture replicas. This means that virtually no model is wearing a finish ever even remotely dictated by the Navy Board. With Navy Board orders on how ships of this time should be painted it is possible to derive a possible authentic scheme, but be aware that it can never be known for certain how *Bounty* appeared, but it is possible to rule out what she did *not* look like.

The aftermarket accessories to be used are from HiSModel from the Czech Republic. For Airfix's *Bounty* they offer a laser cut deck overlay in oak veneer with the correct planking

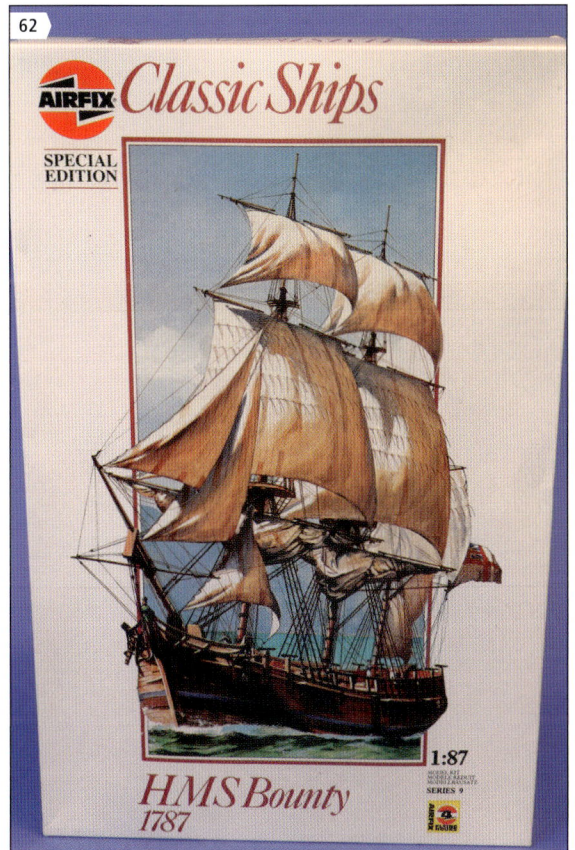

62

The *Bounty* kit. The natural wood payed hull scheme shown on the box is highly improbable and should not be followed. *Bounty* did not fly the White Ensign either, but a Red Ensign.

detail taken from McKay's drawings. The overlay fits precisely over the plastic deck, and also includes scale wooden gratings to fit into or over the hatch coamings. A real star item is a set of CNC cut and sewn sails from a very fine weave cloth with stitched detail. The fabric is pre-dyed a realistic beige colour and every sail that *Bounty* carried is provided. With these sails (or even without them) HiSModels offers a *Bounty* specific set of rigging wooden blocks, deadeyes, and threads. In this small scale, many of the double and single blocks are tiny but perfectly milled to the correct shape with authentic sheave detail. The rigging line set comprises five thicknesses of black and four of tan thread to scale for the model. To help the modeller choose the right block and line, HiSModel has produced a series of rigging guides based on the McKay drawings showing where to use a specific size and shape block and thread. These guides are found on their website and are very comprehensive. The 4-pounder guns are available as turned brass barrels items with laser cut carriages as an alternative to the undersized kit guns. Several pre-blackened fittings such as eyebolts, hooks, and an extensive *parrel* set (a set of rollers to help pull the main yards up the mast) provides essential detail. Accurate cloth flags and pennants are available to finish off your model.

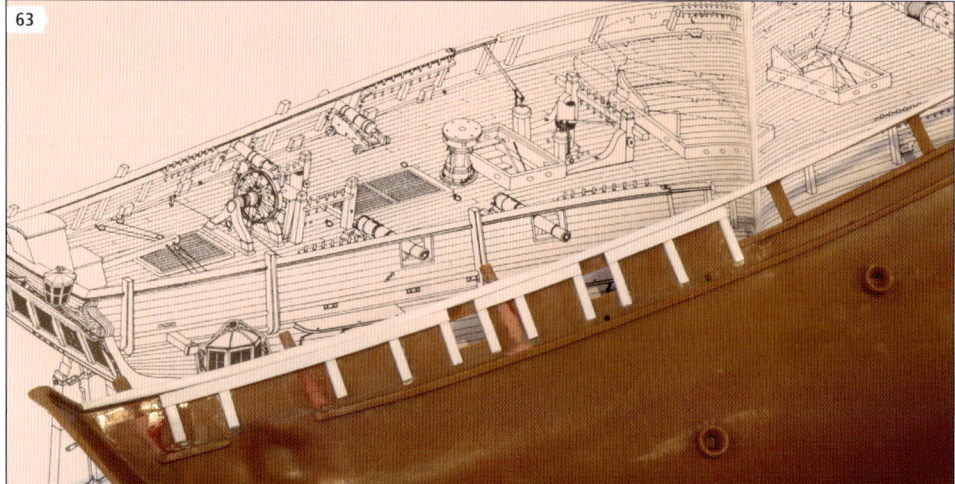

63

The build begins by adding details from styrene strip to the kit's plain inner bulwarks to match the McKay drawings (shown in background). The kit is 1/87 and the McKay drawings are to 1/96, so multiply measurements from the McKay drawings by 1.20.

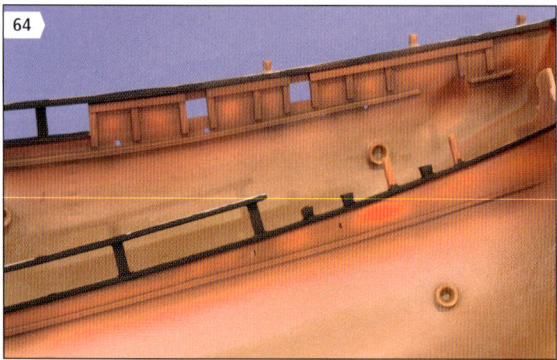

64

The interior is pre-shaded with black and painted red ochre allowing some of the black to show through. The red ochre was highlighted with scarlet sprayed in the centre of each section. A thin wash of burnt umber and black artist's oil colour mixed with odourless thinner colours give additional depth to the finish while blending in all the different shades.

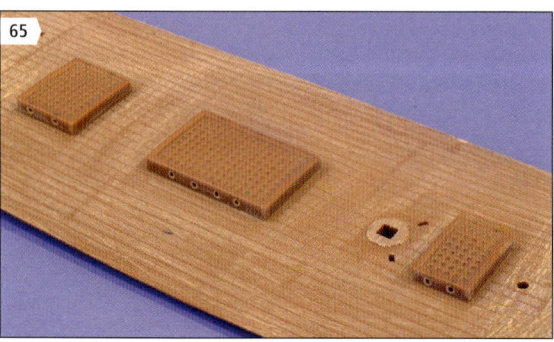

65

The deck moulding is roughed up with coarse sandpaper to take the wooden overlay from HiSModel. The raised coamings were detailed with the missing vent holes drilled into the sides. The holes in the side hatch coamings were to allow additional air into the ship for the benefit of the breadfruit plants. On the real ship, ventilation holes were made in the front and rear coaming sides as well. However, it was impossible to get my drill bit in place to drill these holes out so they were omitted. In retrospect, the holes could have been represented by punched discs of black decal sheet applied to the fore and aft coaming sides. Live and learn!

66

The wood veneer overlay is glued down using slow setting 15-minute epoxy. If any epoxy gets squeezed out of the joints or on the wood overlay, isopropyl alcohol can be used to cleanly wipe it away. Alcohol only works to dissolve uncured epoxy, and any cured glue can only be cleaned away by sanding – which can damage the finish and detail of the wood overlay. Plenty of clamps ensure the entire surface of the overlay and plastic deck is in contact. The deck hatches were pre-painted a dark grey before fitting the overlay to ensure a crisp finish with the deck.

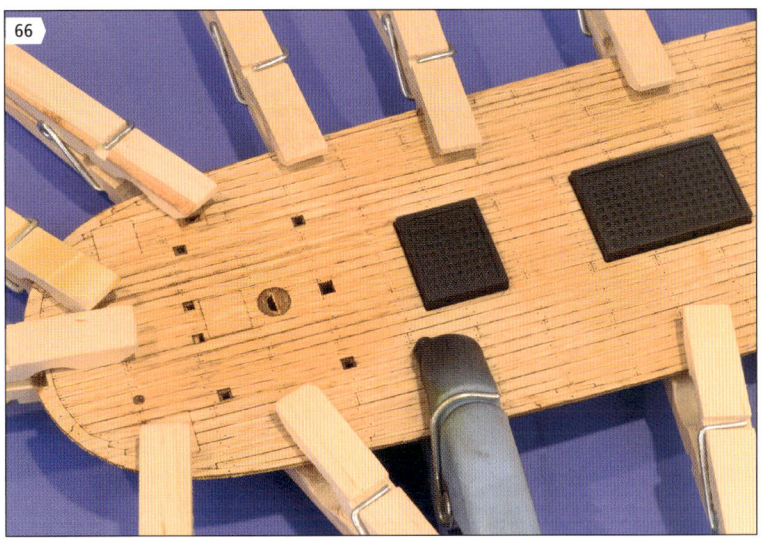

67

The hull and deck were assembled per kit instructions. The underwater hull was painted in copper paint and given a dark patina as described in the previous chapter. Preshading and highlighting was carried out for all the different colours of paint, followed by light washes of burnt umber and black oil colours to create depth and define detail. *Bounty* was originally a merchant collier bought into service and modified to naval standards for her breadfruit journey. As such, the paint scheme applied to the model was based on the Navy Board orders for painting of ship's at the time of *Bounty*. The full details of this research (and educated guesses) are fully described in this author's *Bounty: HM Armed Transport, 1787* (ShipCraft 30). It is unclear if *Bounty* carried any decoration and that supplied with the kit was retained and painted. It would not be incorrect if you choose to sand it off. What appears on the *Bounty* kit was likely taken by Airfix's toolmakers from the replica ship made for motion pictures.

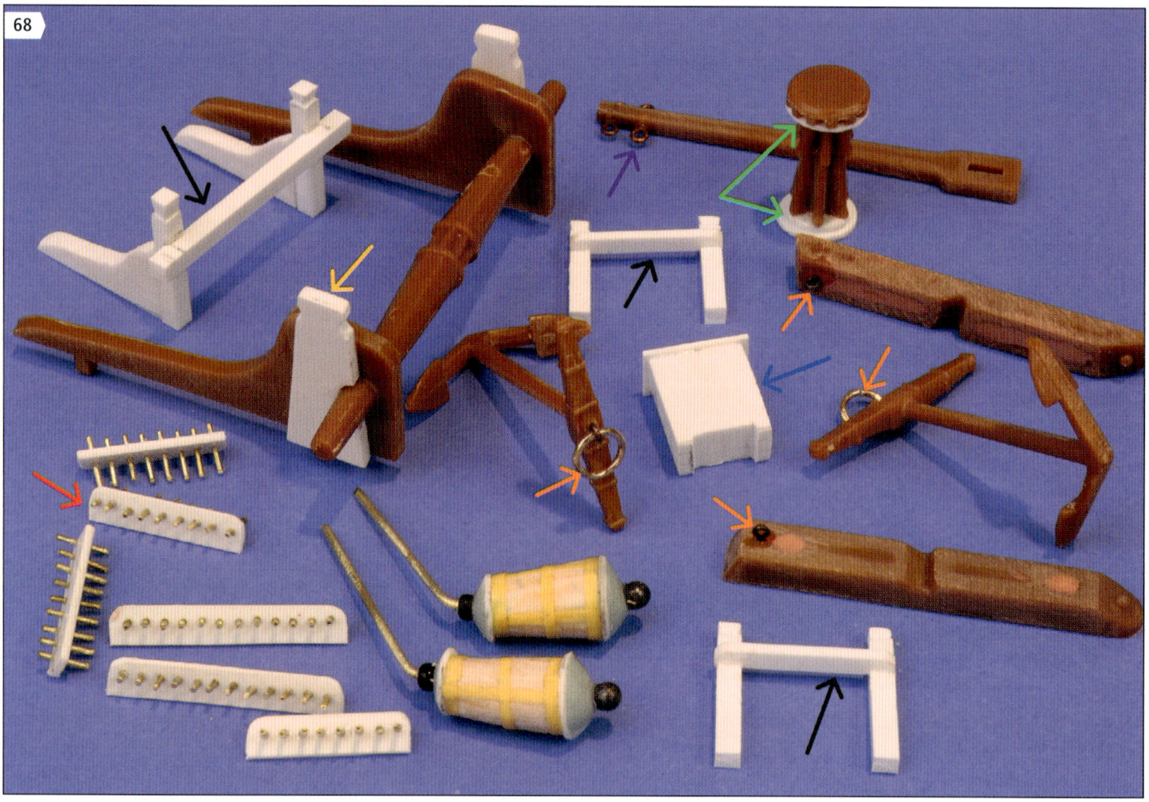

50 SAILING SHIPS FROM PLASTIC KITS

68

Missing deck fittings such as bitts (black arrows), belaying pins and racks (red arrow), and a compass binnacle (blue arrow) were made from styrene using McKay's plans as a guide. The belaying pin rack is a strip of styrene drilled through with a row of holes. The belaying pins themselves are brass rod glued into them and trimmed to length. Kit parts such as the capstan was corrected by adding missing details (green arrows), and the windlass received additional cheeks (yellow arrow). The *tiller* (a lever used to turn the rudder) had eyelets for tiller ropes added (purple arrow). The anchors received a ring from brass wire, and the *catheads* (a projecting piece of timber near the bow of the ship to which the anchor was hoisted and secured) received eyelets (orange arrows).

The kit does not provide stern lanterns and they were made from scratch (bottom centre). The body of the lanterns were shaped from a scrap of balsa that was sealed with thin CA. The frames were added from strips of masking tape. The lantern top is a disc of styrene to which a blob of filler was added and sanded to give a rounded top, crowned with metal BB gun pellet which just happened to be the right size. The lantern base was made like the top, and a small bead (known as a 'seed bead') glued to the bottom. A length of brass rod was bent and one end was fixed into a hole drilled into the base bottom through the seed bead.

Ideally, the lantern body should have been shaped from a piece of clear plastic sprue taken from an airplane kit but none to the right diameter was to hand. Instead scrap balsa was used because it was easy to shape and the whole lantern painted black. Many modellers and kit instructions suggest painting the lanterns gold or some other bright metallic colour. This would have been very unlikely for *Bounty* whose origin was an unglamorous collier whose metal work was all blackened to protect it from the elements.

69

Painted deck fittings are added to the deck. Also shown are the wooden grating and hatch overlays glued into place.

70

Painted deck fittings are added to the deck. The binnacle and ship's wheel was given a deep mahogany red woodgrain finish. All of the fittings were shaded with washes to give them depth. The larger fittings, like the windlass, received airbrushed shades and highlights of the base colour, and then an overall wash of brown and black.

71

The rudder tiller is rigged following the diagrams in McKay's book. Eyelets (red arrows) were glued into drilled holes in the deck and are used like rigging blocks through which the rigging is *rove* (threaded). To complete the illusion, a disc of brown paper was glued on top (white arrow) to simulate a block. A blob of PVA glue painted brown after drying would work as well.

72

The kit guns are undernourished and they were replaced with HiSModel turned brass barrels and laser cut carriages.

73

The guns were fixed to the carriages which were detailed with eyelets and breeching ropes attached. If desired, you could fit all of the tackle used to haul and train the guns for the ultimate detail. The deck has had several eyelets (eyebolts) fitted for rigging following McKay's book. If opting for a simplified rigging setup, even if lines are not ultimately rigged to them they add authentic clutter to the deck that is visually appealing.

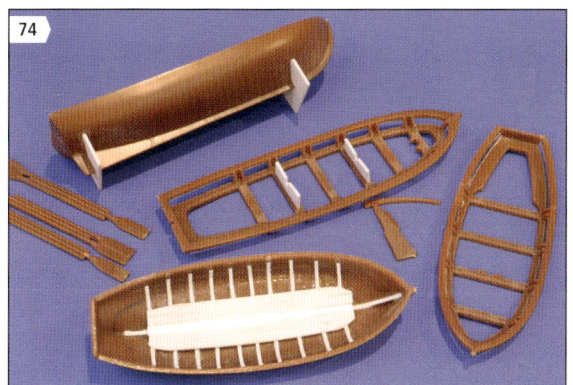

74

The ship's launch and cutter were detailed with cradles, and representative ribs and floors from plastic strip. Remember to fit only the two largest boats included in the kit (the launch and cutter). The smallest boat called the 'jolly boat' was ordered and delivered to the *Bounty*, but was left behind when the ship embarked on her trip to Otaheite.

75

The painted boats were attached to the hull and tied down with thread lashings fed through eyebolts set into the deck.

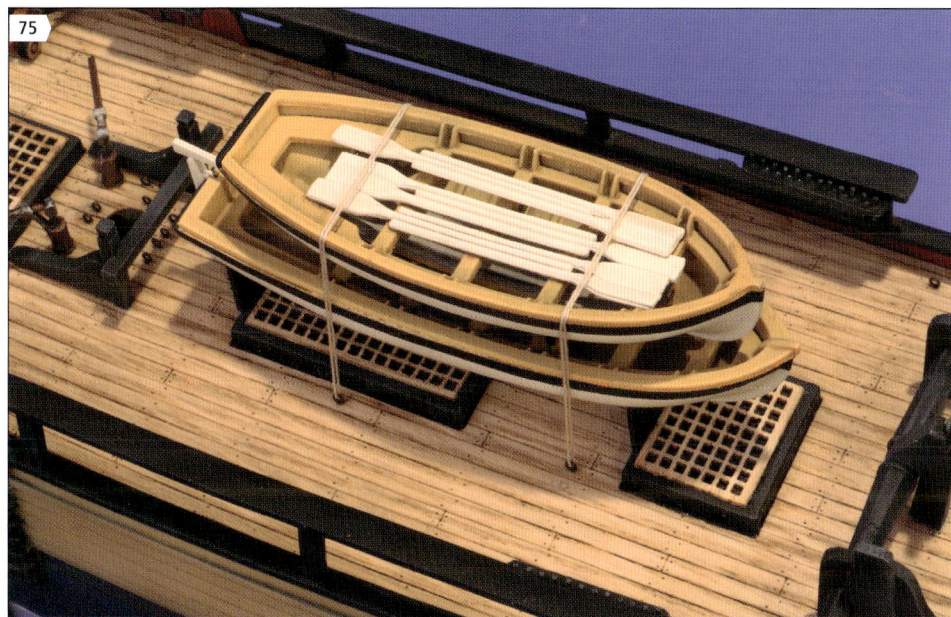

76

The completed hull ready for rigging.

77
The kit's unrealistic deadeye parts.

78
A semi-prototypical way of fitting deadeyes was used to simplify construction without sacrificing authenticity. The wooden deadeyes are the ones used by wooden ship modellers. The sizes purchased were determined using McKay's drawings. The first step is to strop wood deadeyes by wrapping brass wire around them and twisting the tails tight as shown. The kit's moulded deadeyes were removed from the channels, but the moulded chainplates below the channels retained.

79
Holes were drilled into the channels for the new deadeyes. The tails of stropped deadeyes were trimmed and glued into the holes. Any protruding tail under the channel was trimmed flush. Eyebolts were also set into the channels following McKay's drawings.

80
The deadeyes and channels were painted black and glued to the hull. The result is far more realistic than the kit parts and it was not all that hard to do.

81

Mast construction began with making the gratten tops. After sanding the moulded detail off the upper face of each top, a simple lattice pattern was laid out using self-adhesive vinyl tape. The tape is from BECC Model Supplies. Once painted black it will give the impression that the floors were made of gratings.

82

Shown here are the completed mast tops with scratch-built barriers (black arrows) made using brass rod for the stanchions and a strip of styrene for the rail. The masts are detailed with mast bands (white arrow) made from masking tape strips. The mizzen topmast has been replaced with one from dowel shaped using the kit part as a guide (red arrow). Holes have been drilled on each side of the mast tops to take the lower deadeyes (yellow arrow).

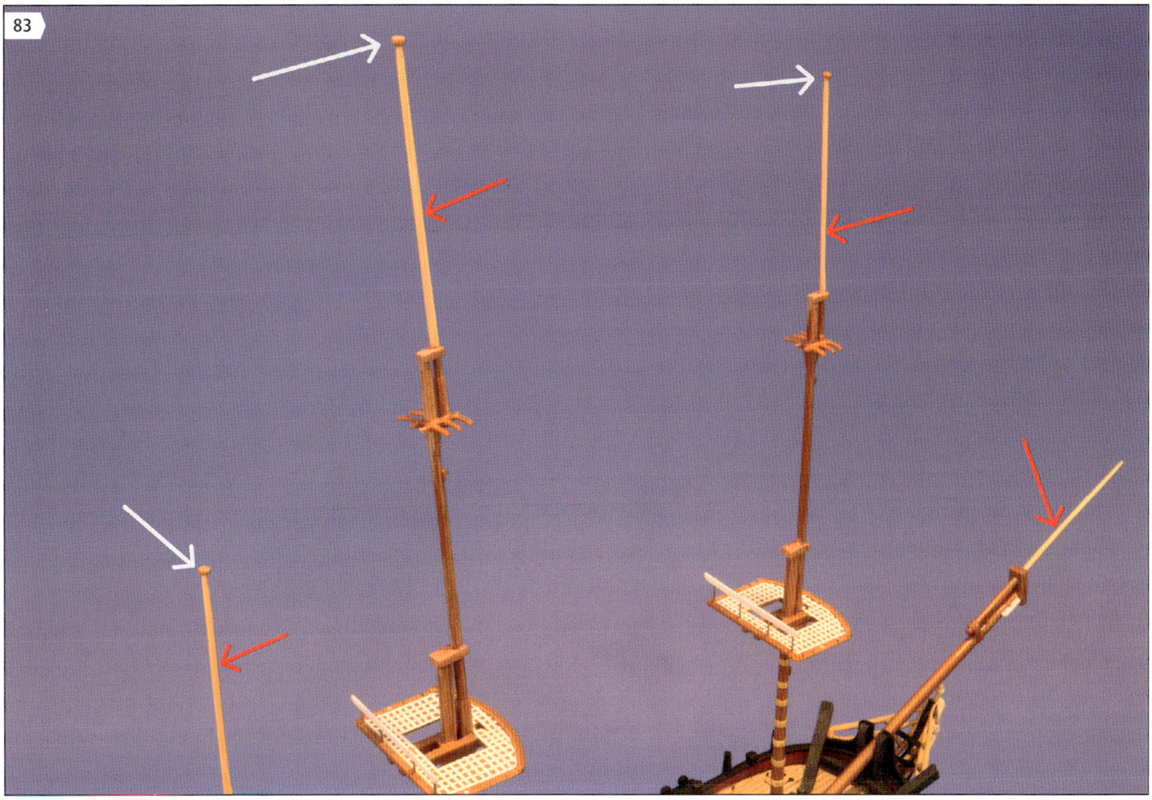

BEYOND THE BOX

83

All three topgallant masts were replaced with ones made from dowel, as was the bowsprit jibboom (red arrows). The mast trucks (white arrows) were cut from the discarded kit parts and glued to the top of the new masts. The masts below the tops were painted with a medium brown woodgrain finish sealed with satin varnish. The mast tops, bands, top and topgallants masts will be painted black.

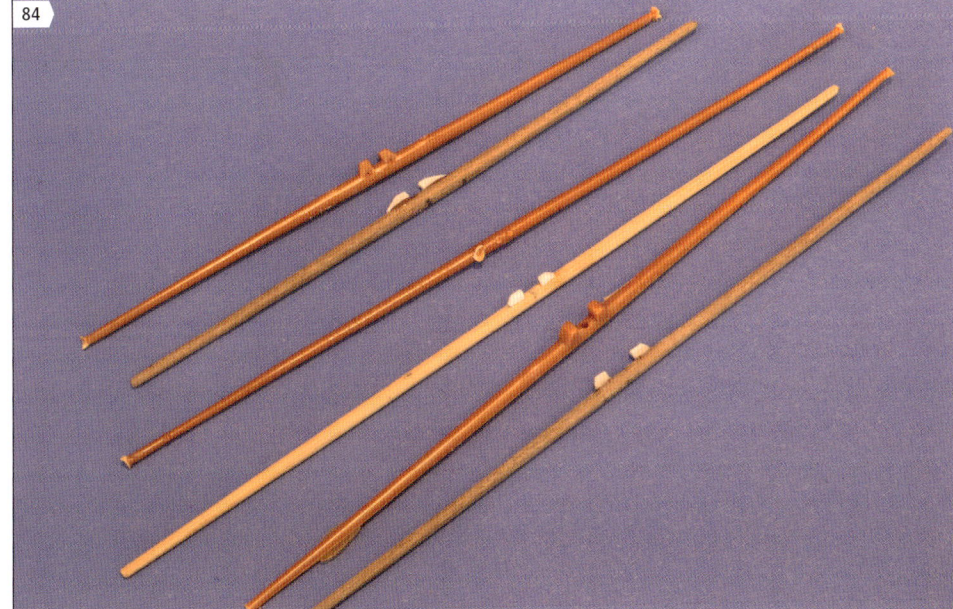

84

The kit's slender and bendy plastic yards with their wooden replacements. They will all be painted black.

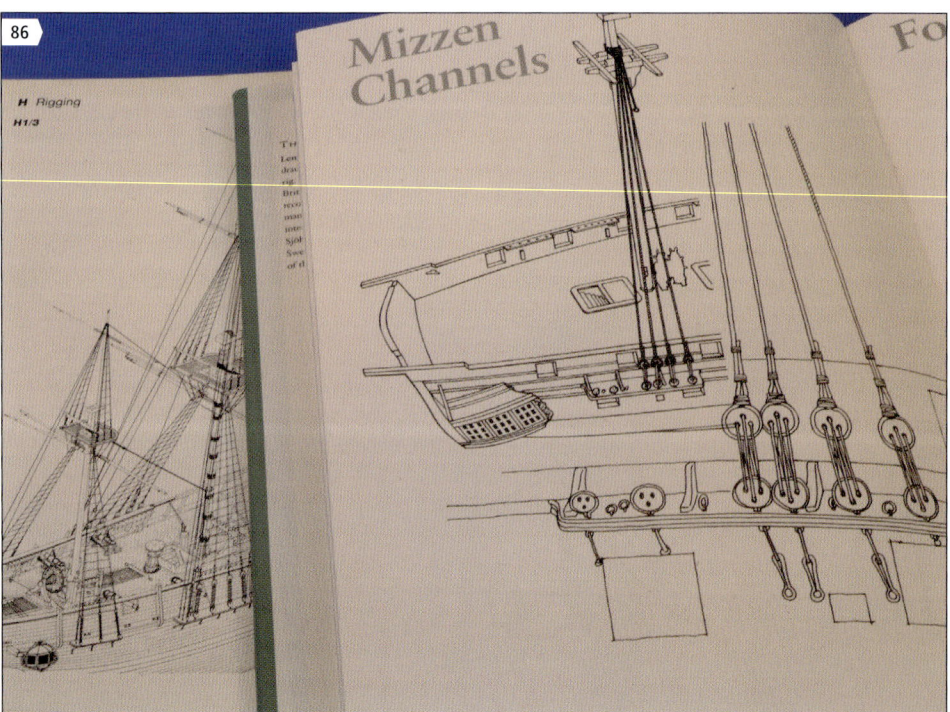

85
The painted masts are fitted into the hull. The effect of shading and highlighting of all the paintwork really provides visual appeal and imparts a period feel to the model.

86
With the hull complete, decisions about the rigging are next. It was beyond my patience to rig the entire ship prototypically, given the small scale and size of the model. There will be some dab hands that can do it, but care must be taken that it does not come out over-scale. For this model the goal was to just set up proper shrouds and reeved deadeyes because they are a prominent feature of the ship. The stays and running rigging will be simplified to the point that it looks purposeful but not chaotic. The illustrations are from Petersson's invaluable book *Rigging Period Ship Models*.

87

The upper deadeyes are shown being stropped and seized to the shrouds. All the knots are fixed with a bit of thinned white glue. When the glue has dried, the excess thread is cut away.

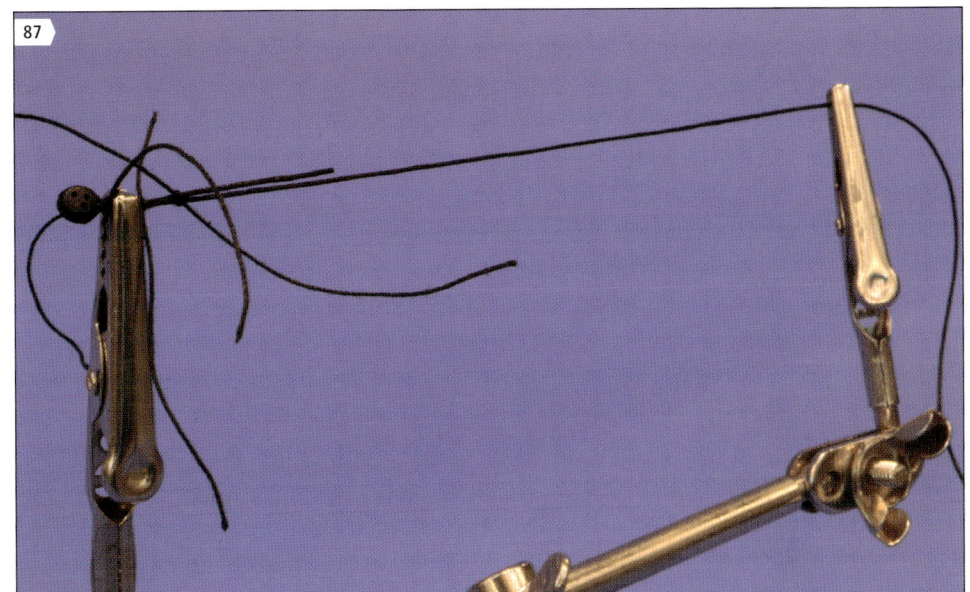

88

After the shrouds have been loosely seized to the top of the masts, the deadeyes are spaced using a bit of brass rod whose ends are bent at correct length. The spacer is inserted into the correct holes in the first member of a deadeye pair (make sure the deadeyes' holes are oriented correctly – they have a tendency to turn within the thread stropping so you could inadvertently place the wire spacer into the wrong hole) while the lanyards on the other member are laced and tensioned. Do not tie off the lanyards yet, but leave them loose. Friction between the thread and the holes will keep the thread from pulling loose. The spacer is removed and the other pair of deadeyes is reeved. The lanyards can now be pulled or loosened to get the correct distance between the upper and lower deadeye, and tension to the shroud. Do not tie off the lanyards until all of the deadeyes on a mast are completed.

89

The lower mast's deadeyes and shrouds are rigged. It is certainly looking better than the kit parts.

90

The lower mast shrouds are rigged and properly tensioned with the deadeye lanyards just like the real ship.

91

The deadeyes and shrouds for the upper masts are shown in the process of rigging. Note how the lower deadeyes are fitted. A brass rod (*sheer pole*) is tied to the shrouds. The lower deadeyes are stropped with thread and the tail of the thread (called a *futtock shroud*) is fed through a hole in the mast top and tied to the brass rod. After all the futtock shrouds are tied off, the brass rod is trimmed and painted black. The shrouds themselves are rigged like the lower mast shrouds.

92

The topgallant shrouds have no deadeyes. Instead, a trestletree is used. On this model, black elastic thread was used for these shrouds. The thread was simply glued to the top of the mast and the tips of the trestletrees and no further action was needed. On a full-sized ship the line is actually reeved through a hole in the trestletree arms and seized to the mast. On the model, the trestletrees were very fine and impossible (for me) to drill through each to properly rig the shroud.

93

The deadeyes and shrouds for the upper masts and topgallant shrouds are rigged next. After the shrouds, the backstays were rigged. The bowsprit was glued in place and its rigging, followed by adding the forestays. What you will find is that when you add the all the shrouds and stays the model's rigging behaves just like it does on the real ship adding tension fore, aft, and laterally to the masts in support of them.

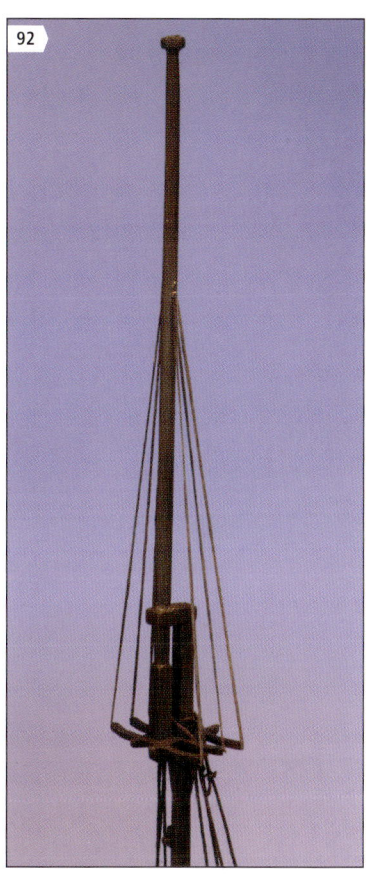

94

The ratlines were added using black elastic thread glued across the shrouds with thick CA. This line is light and easy to manoeuvre and, for this size of model, works well. Thin thread can be used too but elastic thread lets you add some tension to the ratlines so they lay straight across the shrouds without pulling the shrouds out of shape.

95

Bounty's suite of real cloth sails are laid out with the yards they will be attached to. The HiSModel sails are CNC cut and sewn.

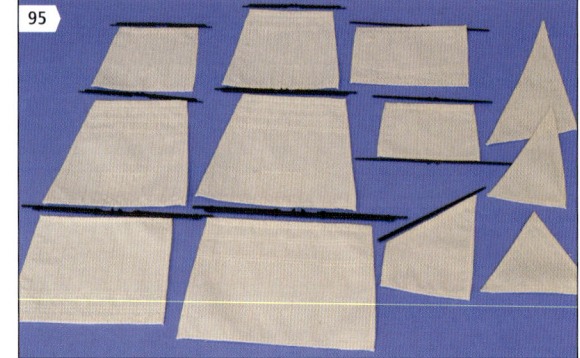

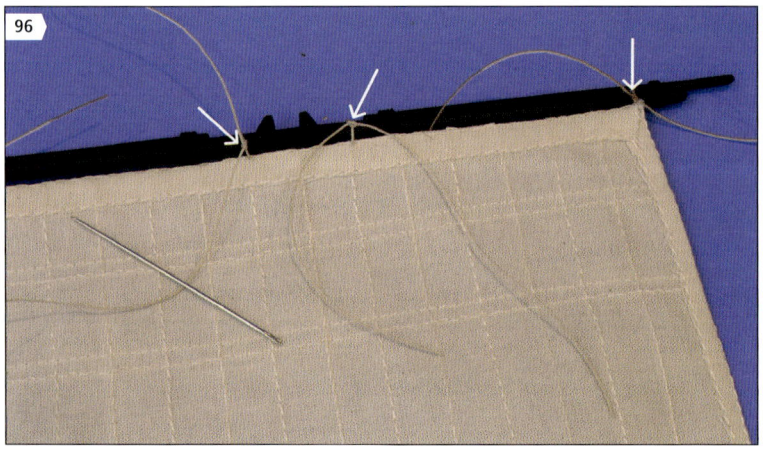

96

The next step is to attach the fabric sails to the yards using a needle and thread for a *roband* (white arrow). For authenticity, the sails should be attached to each yard using evenly spaced robands along its entire length, but only a few were used on this model for simplicity.

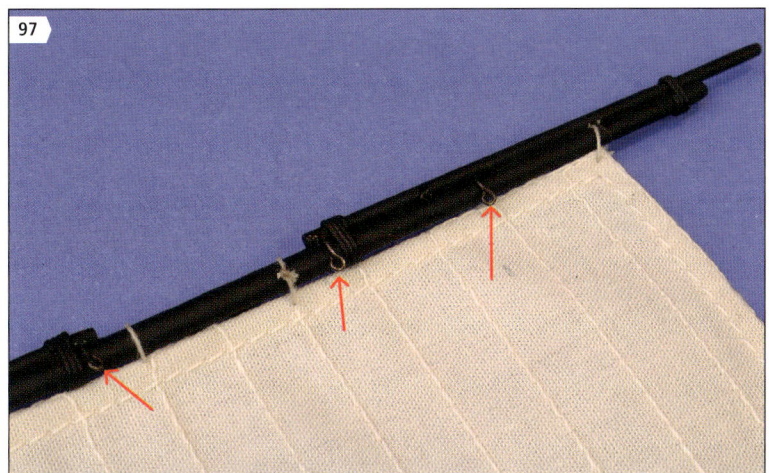

97

Eyelets were mounted to the yards and masts to act like a block would do so lines can be run in a semi-authentic way (red arrows).

98

The yard and sail units were glued to the masts and bunt, clew and leech lines run to handle the sails. *Bunt lines* (ropes that pull the centre section of the sail up), *leech lines* (used to tension a sail to improve its shape to better use the wind), and *clew lines* (one of the ropes or tackles used to raise the lower corners of a square sail, called *clews*, to the upper yard or mast) and run to the pin rails (following McKay's rigging plan). It is interesting to see as you pull on each line what it actually does to the sail, and you can get a real sense of how a full-sized sail was managed.

The fore and main courses are being shown in the process of being furled (or let out, depending on your point of view) because the fabric sail is too long, caused by Airfix moulding the main yard's mounting lug not in the fully raised position. If you want to show your model under full sail, you will have to carve away the moulded mounting lugs and smooth the masts so you can mount them in their highest position. The lines were run to the proper belaying pin hooked to the bottom of a pin and glued into place. The excess thread was trimmed away and loops of line (*rope coils*) glued to the top of the rails to complete the illusion.

99
A close up of the fabric sails.

100
A view of the headsails and rigging.

BEYOND THE BOX

100

101

BEYOND THE BOX

101
Final details are added, such as rudder chains, the lanterns and anchors, and a Red Ensign printed out on copy paper. The fabric sails are far better than the vacuformed items and it's hard to tell the model is from a plastic kit. The figure of Captain Bligh (Airfix's figure is instantly recognizable as the actor Charles Laughton who played Bligh in the 1935 movie *Mutiny on the Bounty*) is painted and fixed to his quarterdeck giving a sense of scale. The kit did not include a figure of head mutineer Fletcher Christian. The toolmakers at Airfix must have all been loyal men. A lot more detailing and rigging could have been carried out using the available references, but even with the simplifications the model looks ready for the South Seas on the quest for breadfruit.

102
A rear view of the foremast sails and rigging.

5: Scenic Models
Jim Baumann's *Mary Rose* and *Sir Winston Churchill*
Airfix 1/400 & Aoshima 1/350 Scales

Jim Baumann of Southampton, UK, is a specialist in small-scale ships set is realistic seascapes. Small-scale plastic kits are ideal for scenic models because the size of the seascape base is quite manageable in size. Moreover, when small-scale models are detailed replete with billowing sails, rigging and figures, they take on a jewel-like quality like no other style of model. With scenic models, special attention must be paid to how the hull, sails and rigging are portrayed because they are central in setting the mood and scene. It is these subtle details that make all the difference between seeing a ship at sea or just a model on a base. To capture this essence you will have to embrace your artistic side and 'feel' the wind as you shape the sails, flags and pennants, and rigging.

There are so many ways to make a seascape base, and the method will make a difference in how well you set the mood. The most common method used by modellers is a layer of plaster on a base with the waves sculpted with a finger. This method works well for calm and flat seas, but when you use your finger to sculpt in the texture of a more lively sea, it is difficult to model the effect of the wind on the water's surface. More often than not, the plaster looks like a child's finger painting that will be painted blue. Professional miniaturists will carve the sea into a thick block of wood like jelutong or basswood. With a large wood gouge, the swells are first worked into the surface of the wood, followed by the wave texture carved in using a small gouge. The advantage of this method is that absolute control is exercised in creating the sea and portraying the effect of the wind. It does take carving experience to do properly and it is a slow process. Another popular method among plastic kit builders is to use one of the many speciality acrylic products to build the sea out of tinted resin. A shallow mould box is made from plastic sheet and the clear resin tinted blue and green is poured in and allowed to cure. The ocean's surface is modelled with the use of thick gels and mediums that dry clear. Some creative modellers make very deep mould boxes so a model of a whale or other sea life can be displayed in its underwater habitat with the ship sailing above. The seas produced using these products can be very eye-catching, but they look artificial and plastic, akin to an *object d'art*.

In contrast, Jim Baumann uses a 'built up' method to make his realistic seascapes that provides a good balance of ease and control for the home hobbyist. In this method, the swells are roughed out using formers - strips of wood, dowel, or scrunched up paper that are glued to the base, again taking care to take into account the strength and direction of the wind. The formers are easily moved around and test fitted on the base until the desired effect is obtained. Over these formers and the base, Baumann's method lays watercolour paper which is then tack glued it to the top of the formers delineating the height of the swell and to the base to create the swell's trough. The paper is left to undulate on its own so it is not unduly forced into any trough. Once swells have been formed, he locks the shape in place by coating the paper with thin CA glue. The CA soaks into the paper creating a hard surface that can take a coat of acrylic gel medium to form the waves. The raised edges of the swells at the base edges are filled with two-part automotive filler (*eg*, Isopon in the UK or Bondo in North America). Before forming the waves, it is important to do some research into the sea your ship is sailing in. What do the waves look like as the wind blows across it? What state of the sea and the strength of the wind will your ship be sailing?

Another consideration to bear in mind is that not all seas are blue. Is the colour grey-green of the Atlantic, perhaps brown and muddy like the River Medway, or the cerulean blue of the South Pacific? The ocean reflects the colour of the sky at the time and season, but its colour is also influenced by how much silt is being churned up into the water, plankton blooms, depth of the water at a particular point, the day's weather, and ocean salinity; all will affect its appearance. If you can, a simple walk by the seaside is as good a reference as any, or colour photos of the ocean of interest. Do not try to reproduce an artist's painting of the ocean because being a two-dimensional representation of the sea the artist must use dramatic shades to reproduce the effect of the sun's light and to create depth. A model seascape is in three dimensions and benefits from the effects of natural light hitting the model and seascape in the real world.

Once a seascape has been created, the next question is what would a ship in this sea look like? What sails would be set in this weather? How are the boats and anchors stowed? How full would the sails be in this wind? What hatches would be open? How would the guns be stowed? What would be the effect of the wind on the shape of the rigging and flags? And for the seascape, what would the shape of the wake be like given the shape of the hull, and how much froth should be modelled to accurately portray the speed of the ship? This is important because, sailors used the state of the seas and wake to judge how well their ship was cutting through the water and her speed. These are questions of seamanship and how a ship should appear at sea can be a mystery, especially for those who have not been on a ship at sea before. An excellent reference is John Harland's 1984 *Seamanship in the Age of Sail*. His book is exquisitely illustrated by Mark Myers, who had extensive seagoing experience in sailing ships, and shows important seamanship skills like how to coil rope around a windless, the setting of sails, reefs, how anchors are handled, and different sets of sails to suit the prevailing weather conditions, for example. This book will help you get all the details right, and if you add scale figures (and you should), your miniature matelots will actually be doing something as opposed to walking the deck like a zombie. This reference will give you good technical skills and knowledge, but once again you will have to embrace your artistic side to make it all come together. Following Jim Baumann's approach to modelling will amply illustrate what I mean.

MARY ROSE

The first scenic model is *Mary Rose*. The *Mary Rose* was launched in 1511 under the patronage of Henry VIII, and she served for 33 years in wars against France, Scotland, and Brittany. Her last action and sinking was in 1535 against a French invasion fleet in the Solent, off the Isle of Wight. After discharging a broadside, she turned and was caught by a gust of wind that pushed the still open lower gun ports under water. The hull quickly filled with water and she sank. Her wreck was raised in 1982 and can be visited at the *Mary Rose* Museum located in Portsmouth Historic Dockyard. The museum is well worth a visit because it displays far more than just the ship timbers, including a large number of exhibits on life at sea on a Tudor warship and forensic reconstructions and identification of specific crew members and officers based on skeletal remains and personal items.

103

The kit sports a full hull and is designed for easy assembly. A feature of Tudor era warships was extensive anti-boarding netting that stretched over the main deck (red arrow), but because Airfix moulded the netting as a single solid part it was discarded and real netting installed instead. The simplified and undetailed main deck part (yellow arrow) was also discarded because with the installation of real netting, the main deck will be visible requiring a properly fitted-out main deck to be constructed from scratch.

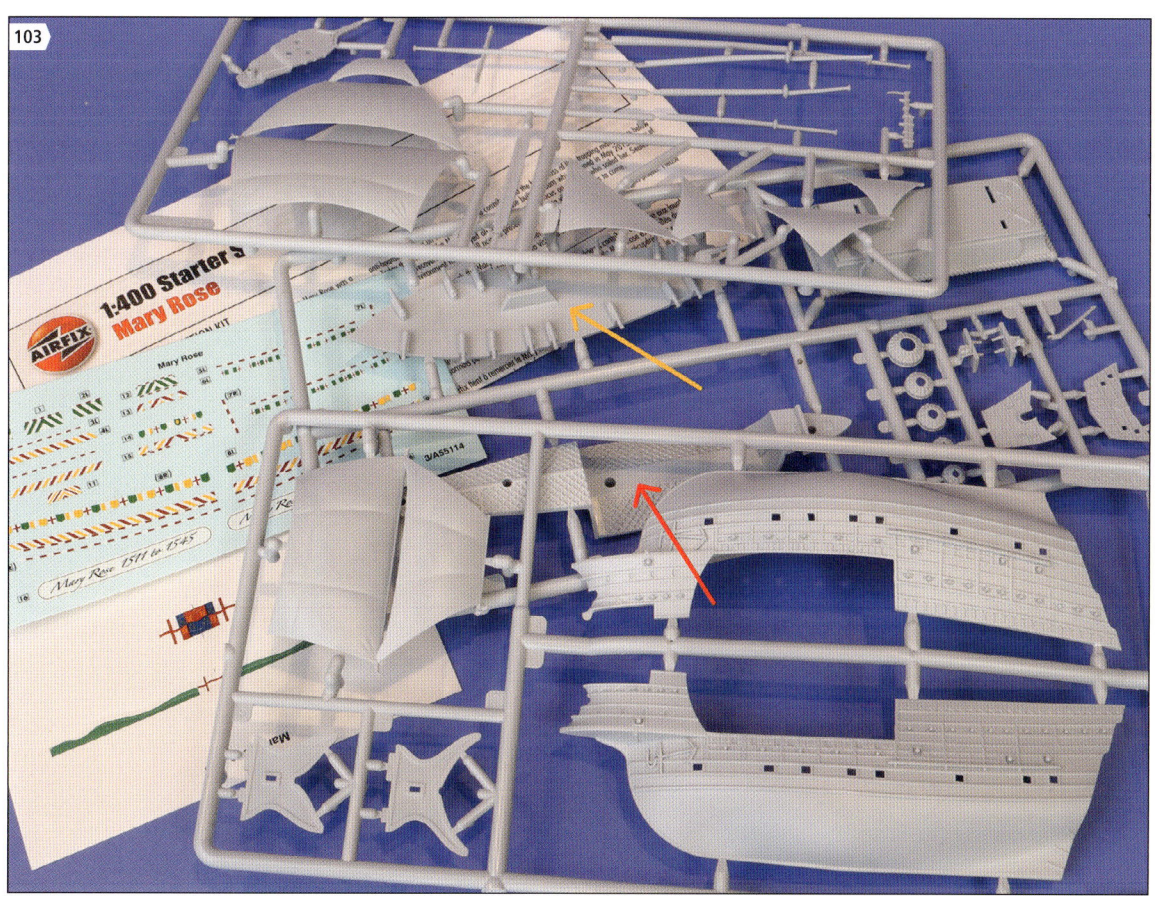

104

Before assembling the hull, the inner bulwarks were sanded to a more scale thickness by scraping with a knife and sanding sticks. Hull apertures, such as the archer blinds were opened up to give a more delicate and realistic appearance. The ScaleWarships etched set (see Figure 28) provided new decks, shrouds and ratlines. The photo shows the hull was assembled and painted using different shades of brown. (*All photos in this chapter used with permission from Jim Baumann*)

105

With the hull assembled, the next step is to make the seascape base. The base was made from wood with a hole cut out to accommodate the lower hull. Swell details were roughed onto the base by gluing round toothpicks taking into account the direction of the wind. The hull was dry-fitted into the hole at different angles to test different degrees of heel caused by the force of the wind against the hull and into the sails.

Several angles were tested to get the right effect when the hull is viewed from all sides. The base was coated with layers of plaster to build up the final seascape with swells. Products like artist's gesso can also be used, especially for the final layers to seal the plaster and 'smooth' the sea in preparation for painting.

The sea was painted using a mix of grey, blue and green enamels, and the key to a realistic sea is to subtly shade the troughs and swells, and have the shades a gradually change colour. Too often when painting a sea a modeller paints in dramatic differences in colour that just does not look realistic. Jim Baumann achieves subtlety between his shades and highlights using a 'wet-on-wet method' with enamel paints. A base colour is mixed and painted to the base. Whilst the paint is still wet, the different shades of highlight and shade are daubed on then brushed and blended using a lot of paint thinner. With enamels, even if the paint has started to dry out, it can be rewetted and the colours blended together with thinner.

Each hull shape creates a different wake pattern. Examination of photos of different shaped hulls at sea will give you a good indication as to the one you should be modelling around your model. Some ships with slender narrow hulls cut through the water, while bluff bowed ships tend to push through the water leaving a different, often more turbulent wake. The wake around *Mary Rose* was built up using acrylic texture gels and touched in with white paint. As the wake extends out from the hull, it takes on a diminishing frothy white appearance. This frothy appearance was added using a 'dry-on-wet' technique. The grey-green paint is allowed to dry hard and then some thick white paint is daubed where the wake should be. A copious amount of thinner is used to blend the white paint into the grey-green, drawing out the white paint using a brush. The thinner essentially renders the white paint transparent as it spreads into the grey-green base.

Acrylic paints do not really lend themselves to this kind of wet-on-wet or dry-on-wet blending because they are fast drying and cannot be reconstituted with thinner. Acrylic thinner tends to dissolve paint in chunks and is impossible to blend. Instead, the shades will have to be applied in several thin coats, with each shade being just a little darker or lighter than the one preceding it. Several, almost transparent coats would have to be applied to build up the changes in shade. The shades are blended together using thin transparent washes of the base colour mixed with water and acrylic medium over the seascape. Clear gel mediums can also be used to add depth, and whitecaps and the ship's wake modelled using acrylic texture mediums and coloured with touches of off-white paint.

SCENIC MODELS

106

With the seascape base roughed out, attention returned to the hull. The hull was detailed with doors with curved tops seen on contemporary illustrations of *Mary Rose*. These doors found on a left-over sheet of photoetched brass designed for modern warships were used because the shape and size was about right. The hull bottom was painted a very dark brown-black to represent the pitch composite used to protect the underwater hull. The kit provided decals for the decoration but there was not quite enough to properly cover the whole area. A second kit was procured for the decals to make up the difference.

107

At this point of the build, the kit was compared to the latest research on the real *Mary Rose* and it was discovered that the kit's stern and forecastle decks sit too low inside the hull. The error was likely caused by Airfix basing their model on the newly excavated *Mary Rose* where those decks had deteriorated over the past 500 years under water and were not known to exist. Research on the wreck years later eventually determined that these decks existed and where they were located. The missing decks were made by first creating a paper template (red arrow) to determine the correct shape (stern castle shown here). Its location on the model was estimated by a 1/400 figure (blue arrow). The figure's head should just come to the top of the bulwark. The same process was repeated for the newly discovered forecastle deck.

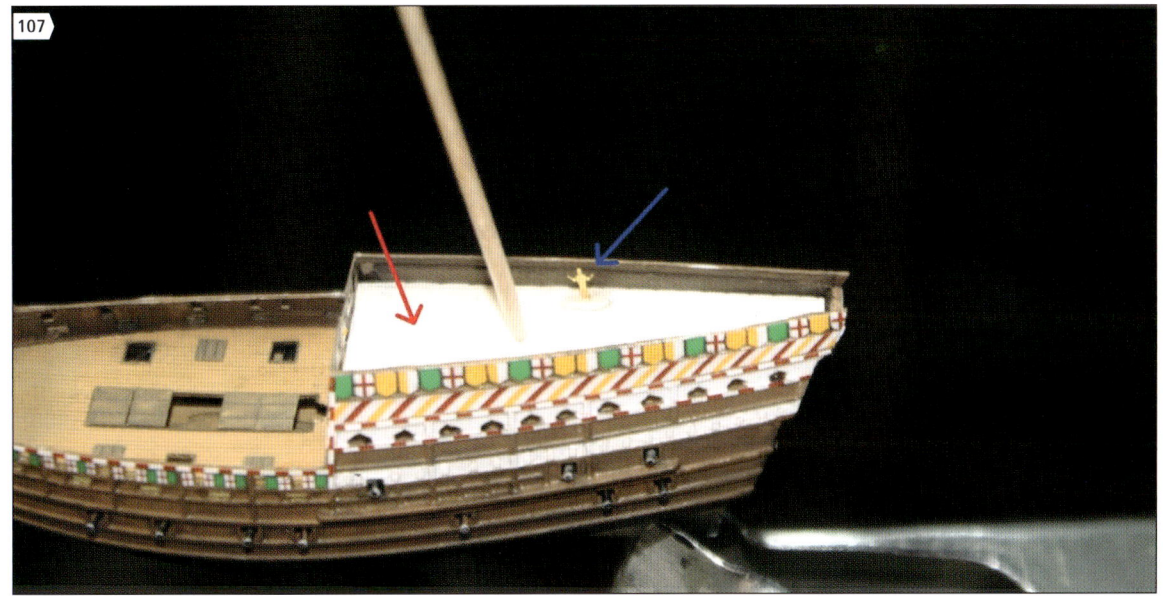

108
The new decks (stern castle deck shown here) were cut from scribed plastic sheet and holes drilled for the masts. The wood grain finish was achieved by painting the deck a light brown colour and building up the variation in colour with several thin washes. The washes were concentrated in different areas to create a natural looking patchwork. The same technique was used on all wooden parts of the ship.

109
A prominent feature of *Mary Rose* is her anti-boarding nets. The first task is to build the framework to hold the netting from plastic strip and brass rod. The model itself is very small, and to create a jewel-like miniature, it is important to keep all of the scratch-built components to scale.

SCENIC MODELS

110
The netting frames are painted brown and figures are added to the decks. The figures are generic plastic and PE brass items used to crew modern warship models. The figures were painted in Tudor clothing so plenty of white shirts and leather jerkins. The masts were built and test fitted into the hull to precisely determine where the netting will have to be cut.

111
Finding the right netting involved quite an extensive search with fabric tulle, teabag fabric, and fine wire meshes all experimented with. The big challenge was to find a mesh pattern (*eg*, diamond) that best resembles that used on *Mary Rose*, and that the holes are not too large for a scale enemy sailor to get through, nor too small that is obscures the deck and details of the model.

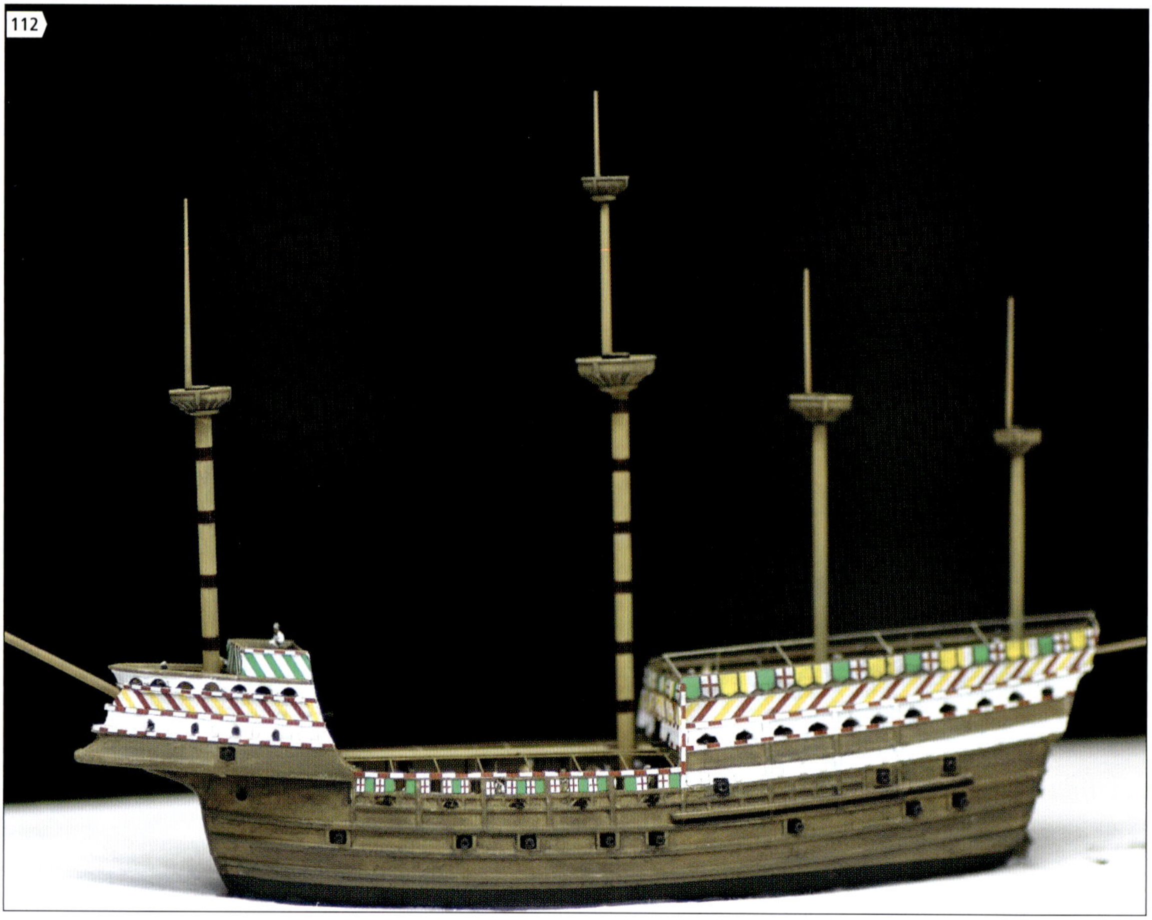

112

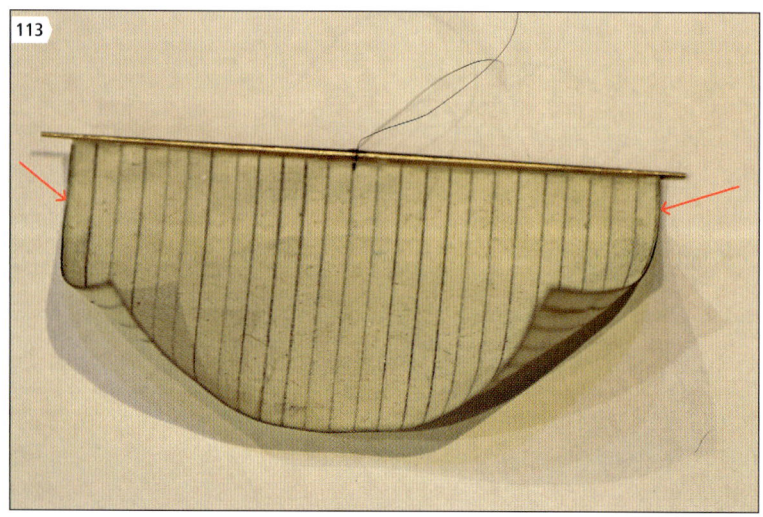

112
The kits masts were remade from brass tube and the visible edges of the crow's nests were ground down and sanded to a more scale-like thickness. The masts and bowsprit were painted and detailed with *woldings* (mast bands made from rope instead of iron) made from black decal strip.

113
The sail blanks were cut from fine paper using the kit parts as a guide. The blanks were detailed with seams drawn in with pencil. The sails were soaked in water and laid over the kit sails as a mould and gently dried in an oven. *Boltropes* (a strong rope stitched to the edge of the sails to strengthen it) around the sails were replicated with wire glued to the edges (red arrow). The wire will help the sails keep their shape and allow some adjustment of the billowing effect. The sails were painted with several thin washes of different shades of tan paint. The yards were turned aftermarket brass items designed for modern steel warships that were the correct diameter. These were overlong and cut down to size. Alternatively, the yards could be easily made by tapering some fine dowel or bamboo skewer.

SCENIC MODELS

114
The ScaleWarship shrouds and ratlines were painted dark brown-black, with a lighter shade touched in with a brush for the ratlines themselves. The etched parts are flat and a blob of white glue was added to each deadeye for depth. Once the glue had dried, the deadeyes were painted a brownish grey and, with a fine pen, three dots were added to represent their reeving holes.

115
All the sails and yards have been added to the masts with a piece of fine wire twisted around the yard's centre, and the tails twisted around the mast at the desired height. The angle and deflection of the yards and billowing effect of the sails was determined by imagining the direction and strength of the wind. The wire boltropes glued to the sails allowed adjustment of the billowing effect and hold the desired shape of the sails.

116

With the sails set, running rigging was added using a combination of heat stretched plastic sprue, elastic rigging thread, and wire. A feature of all of Mr Baumann's miniature models is that small size is no barrier to him fitting as much rigging as possible to the model. Please feel no shame if you leave some lines off because you just cannot get your tweezers in between all the lines or if eyesight is a challenge. It just takes practice and patience.

SCENIC MODELS 77

117

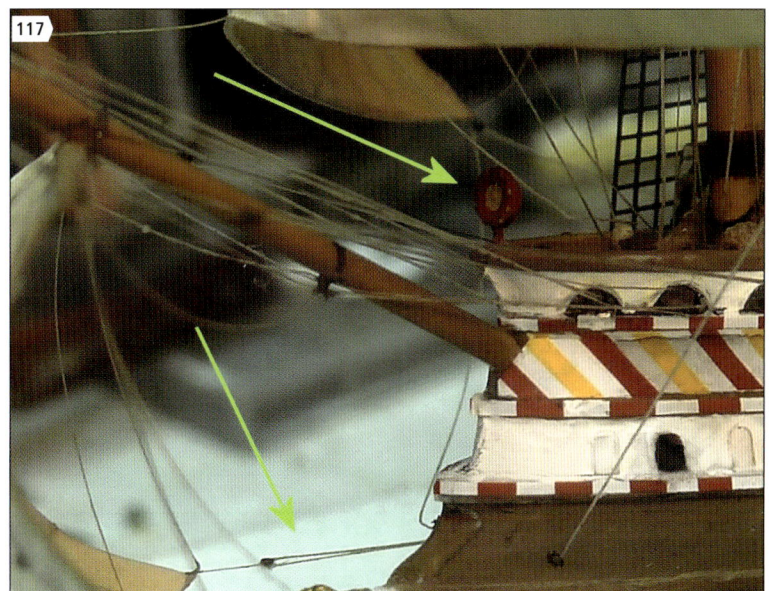

117

Continued research showed that the Airfix kit is missing several small but important details such as the Rose of England and a figurehead of a growling lion head (green arrows). Their small size made it feasible to carve them from scrap styrene. The photo also illustrates several lines used to control the spritsail, all made from painted heat stretched sprue and wire. These materials are ideal because the lines can be shaped over the end of a round paintbrush handle to capture their sag from their weight and the effect of the wind blowing on them. The blocks were made with blobs of glue or discs of paper painted brown.

118

118

A feature of Tudor era warships was grappling hooks (blue arrows) attached to the end of the yards to snag an enemy's rigging so the ship could be drawn in close and boarded. These tiny items were made with wire. The leech lines attached to the front and edge of the sails were made from heat stretched sprue with the blocks punched paper and painted brown (red arrow) for greater fine detail.

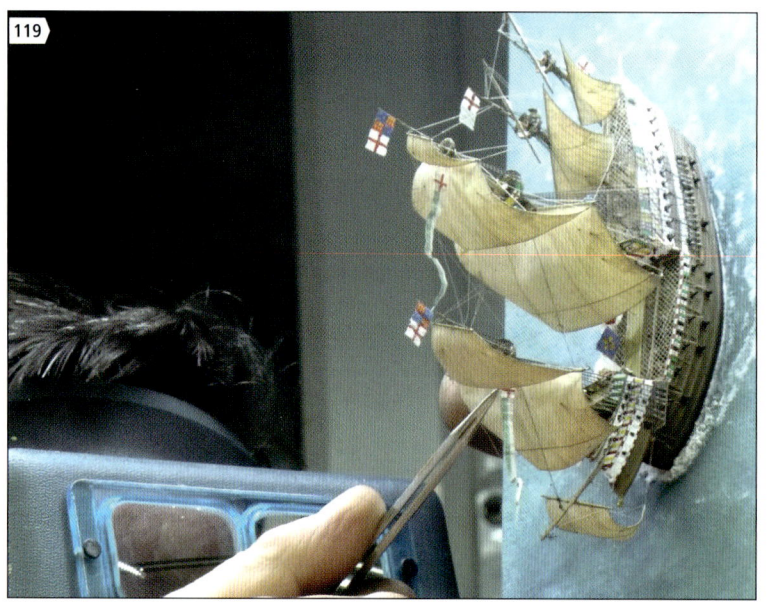

119

Mary Rose being set into the seascape.

120

The model set into the painted seascape. The subtly of the water-surface shading is apparent and the sea state matches the set of the sails. The flags help enliven the scene and in this small scale capturing the delicacy of fabric is no easy job. *Mary Rose*'s flags were researched and printed (also in reverse) onto paper. Copier paper is very thick, and gluing two pieces together would make for a very heavy flag in this scale. The paper was carefully delaminated with a sharp knife by carefully cutting and pulling the layers apart leaving the flag's image on an extremely thin paper backing.

SCENIC MODELS

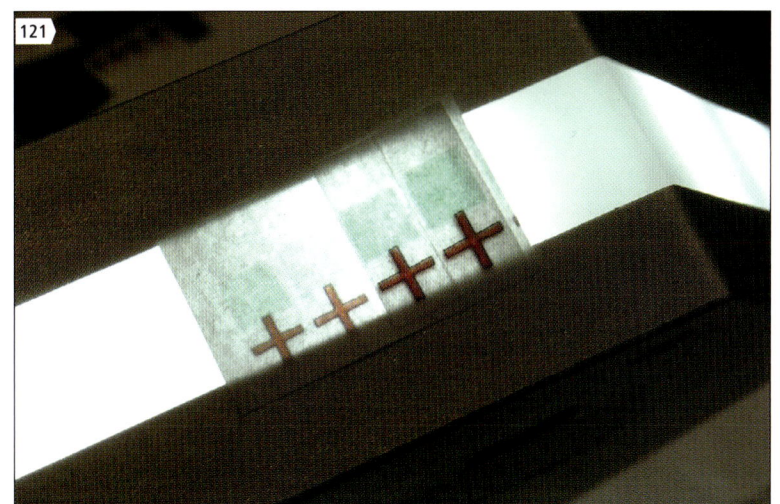

121

The delaminated flags were glued together over a light box to ensure the designs matched up. The paper is so thin that even the minutest misalignment would show through on the other side in the light.

122

The flags were shaped and added to the model, again keeping in mind the strength and direction of the wind. With the model resplendent on its base and a sky backdrop placed in behind for the photo, Jim Baumann really captured the look and feel of this Tudor warship at sea.

The Sail Training Ship *Sir Winston Churchill*

The second example of Jim Baumann's modelling is the schooner *Sir Winston Churchill*. She was designed by Camper & Nicholson and built in 1966 at Hessle in Yorkshire, intended to take youth on sail training trips. A patron of the project was Prince Philip, Duke of Edinburgh who enthusiastically supported and encouraged young people around the Commonwealth to take up sports and outdoor activities. *Sir Winston Churchill* trained youth until the year 2000 and is now a private yacht.

The 1/350 Aoshima kit is very simple, and the mouldings were reworked and detailed like *Mary Rose*. Unlike the *Mary Rose* which is square rigged, the *Sir Winston Churchill* is fore and aft rigged and the challenge was to realistically model her at speed as a greyhound of the sea, and to capture the look of how the sun hits her massive sails which are not canvas but made of lightweight modern synthetic fabrics that has a translucency no other material has.

123

The completed *Sir Winston Churchill* set on the seascape. The size of this 1/350 scale model is tiny compared to Mr Baumann's hand.

124

The base for this seascape is made from watercolour paper laid over wooden formers glued to a wooden base. *Sir Winston Churchill*'s hull is shown being set into a hole in the the seascape and any gaps between the hull and seascape base was filled with acrylic medium and with white PVA glue.

125

Sir Winston Churchill's sails were made from paper using the kit sails as a guide to size and shape. Her actual sails are made of terylene, and this modern material is translucent in nature and seams can be seen through the sails in most lights. To recreate this look, the paper sails were painted with several very thin washes of grey and hemp coloured paint. These washes soaked into the paper making it slightly translucent allowing the pencilled seam lines to be seen through the sails against the light with some subtle shading around each seam. The ship's rigging was wire, elastic thread, and plenty of stretched plastic sprue. Blocks were replicated with blobs of glue or tiny disks of punched paper. On this model, the lines on this fore and aft rigged ship were modelled taut as if under the strain of the wind. Footropes were made from heat stretched sprue, and furled sails made from thin paper scrunched up – not twisted – as if the sail's bottom edge was being pulled up by ropes.

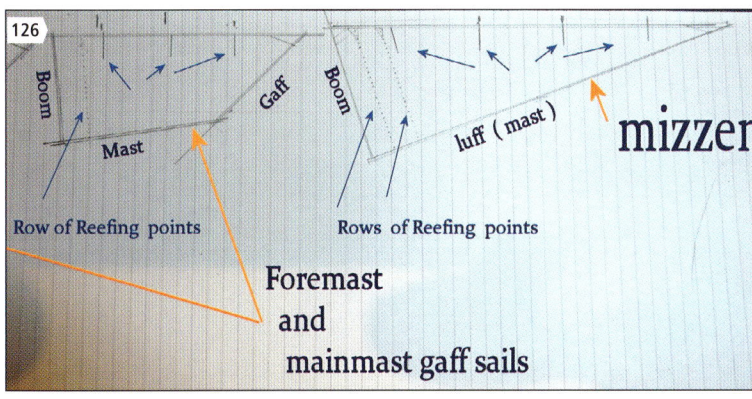

126

This photo shows one of the sail plans used for the model. The various parts of the sail are labelled so that no detail is lost and the correct edges are attached to the correct spar.

127

The sea was carefully modelled with a wake that captures how this hull shape sits and cuts through the water. Careful consideration must be given to the whitecaps added to the wake and seas to give the viewer a sense of the speed the ship is going. Adding too much looks unrealistic and/or the ship is sailing in high seas; too little and the base becomes visually uninteresting and the ship seems to be barely moving despite the set of the sails.

128

A view of the bow wave.

129

A view of the ship's hull and deck. A great deal of true to scale detail was added to the kit parts, including companionway ladders, vents, lights, bitts, and rope coils. Bowsprit netting was from etch brass mesh sheet, and shrouds and ratlines taken from generic sets designed for steel warships.

SCENIC MODELS

130

A profile view of the completed model.

131

A fine port quarter view of the *Sir Winston Churchill* model that evokes the speed and beauty of the actual ship. A detailed build of Jim Baumann's *Mary Rose* and *Sir Winston Churchill* can also be found on the Modelwarships.com website. The build log further discusses his techniques and considerations for a realistic scenic model, and how he recovered from mistakes made along the way.

6: Hybrid Modelling
Victor Yancovitch's *Le Soleil Royal*
Heller 1/100 Scale

Large French ships like *Le Soleil Royal* (The Royal Sun) commissioned during Louis XIV's reign were amongst the most elaborately decorated in maritime history. She participated in the War of the League of Augsburg between France and the Grand Alliance of the Holy Roman Empire, the Dutch Republic, England, Spain and Savoy. Heller's 1/100 *Le Soleil Royal* was released in 1974 and was the largest and most ornate model sailing ship kit at the time. The model features superbly detailed carved work and she is fully rigged like a real ship, including hundreds of plastic blocks in different sizes and shapes that have to be properly stropped with thread. A loom is included to wind the ratlines and shrouds, but given the size and scale of the model, the use of near-prototypical techniques of wooden ship modellers should not be a problem.

The kit is an exacting replica of the 1/40th scale model of *Le Soleil Royal* model displayed in the entrance hall of the Musée de la Marine in Paris. This beautiful model was made in 1839 by the artist Jean-Baptiste Tanneron, but it has a problem with the shape and volume of the underwater hull. It is far too shallow in volume – if the actual ship had adopted these lines she would have been top heavy and prone to capsizing. Most modellers are content knowing that, built out-of-the-box, their model is a model of a model and what is of most interest is the grand baroque style decoration she carries. Modellers who wanted to correct the hull must either cut it away for a large-scale waterline model, or used paint to give the impression that the lower hull is much fuller. The kit's waterline can be ignored and her white painted bottom taken up much higher to the very bottom of the lower wale. This ruse is historically acceptable and can be seen on many model ships and paintings of the period, and the trick really does fool the eye into believing that the lower hull is much deeper in proportion.

A more radical option is to rebuild the hull bottom to a more realistic shape and volume and marry that to the kit's upper hull. Victor Yancovitch of Burns Lake, British Columbia, Canada did just that. Mr Yancovitch has spent a lifetime scratch-building ship models from wood, and some of them to such a large scale that they could be sailed on a lake under full sail. He has used his experience to build new hull framework out of wood with the correct underwater hull shape using a 'plank on bulkhead' approach. Drawings of ships at this time provided only a few sections of the lower hull so the hull had to be reconstructed by extrapolating the known lines to the rest of the hull. However, this approach alone is insufficient because every ship's hull has curves and shapes at the bow where she cut through the water, and the shape also changes towards the stern. These curves, that also determine how well a ship sits and cuts through the water, are not captured by contemporary plans. Several modellers worldwide assisted Mr Yancovitch with this problem by incorporating hull shapes from contemporaries of *Le Soleil Royal*. This effort meant examining models, paintings, and drawings and incorporating the features found there into the new lines until something plausible was created.

The ship's new body lines are traced onto a sheet of plywood and cut out. These become the bulkheads that are

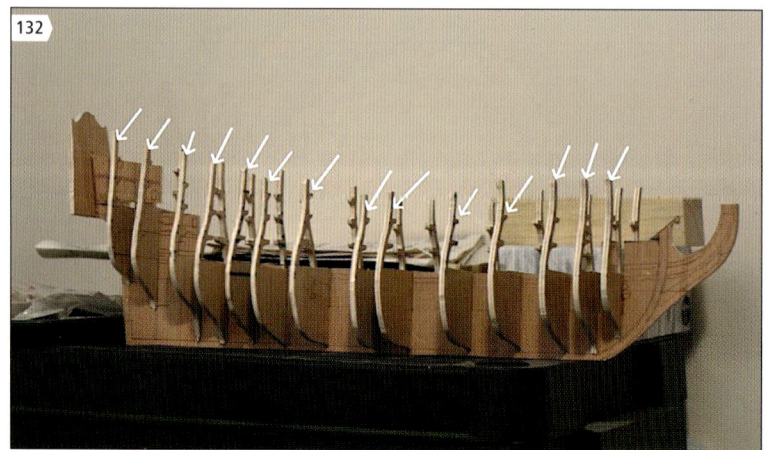

132

The new wooden framework for *Le Soleil Royal*. Just by looking at the bulkheads you can easily see the greatly increased volume of the lower hull. The horizontal extensions on each bulkhead locates each of the ship's decks (white arrow). Eventually, a *cambered* (curved) beam is glued between the extensions to form the deck beam. The edges of the bulkheads are bevelled so the kit parts lay flat across them as the ship tapers at the bow and stern. (*All photos in this chapter used with permission from Victor Yancovitch*)

HYBRID MODELLING

slotted and glued transversely to a profile of the hull to create a new hull framework. The kit's upper bulwark parts were test fitted to the framework and adjustments made to the bulkhead so the plastic parts fit well against them. The edges of the wooden bulkheads had to be bevelled so the plastic parts fully lay against them, and adjusted so that the sheer of the hull's framework matches the plastic kit parts. The area below the bulwarks was then planked over with strips of fine hardwood, just as one would normally do with a wooden ship kit model. How to plank a wooden hull is described in this author's book *Ship Models from the Age of Sail: Building and Enhancing Commercial Kits*. The other advantage of this hybrid approach to modelling is that *Le Soleil Royal*'s hull below the decorated upper bulwarks was payed, and having the hull planked in wood allowed these sections to be varnished and look exactly like a payed finish.

Trying to build a plastic version of the ship's skeleton would be very difficult. The styrene sheet that would have to be used for the bulkheads and hull profile would need to be very thick, and thick styrene is difficult to cut out using the usual 'score and snap method'; moreover, despite its thickness, styrene does not have the same stiffness as wood. Wood is easier to cut and shape, and shaping thin wooden strips to fit the double curvature of the hull (a process called *spiling*) is much more straightforward. Indeed, there are some wooden kit manufacturers who, for this very reason, exclusively use plywood and strip wood to build the hull, with the detail parts made in plastic.

133

A cardboard copy of the kit's upper bulwark parts was used as a template to adjust the new wooden framework to accommodate the kit parts. The cardboard template was coloured to help visualise the different elements of the hull sides and delineate what would be painted and what would be left in a natural wood finish.

134

The main plastic hull parts are fitted to the framework.

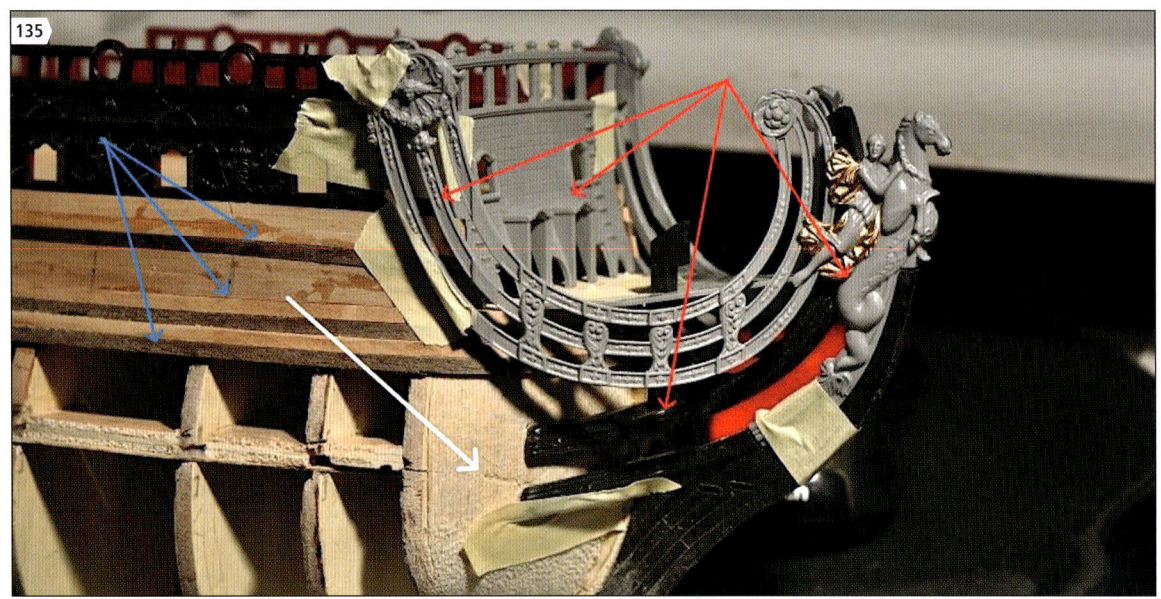

135

A close up of the plastic bow pieces fitted to the new hull. Some of the wood planking (blue arrow) has been applied to the framework to help fit the plastic parts (red arrow) to the hull. A solid wooden block was glued into the front of the framework and shaped to the new bow contours (white arrow). This wooden block will help the wooden planking strips curve neatly around the bow.

136

A close up of the kit stern and side galleries being fitted to the new wooden framework. The transom piece has also been taken from the kit. The guiding principle was that if it was to be painted and/or had decoration, the kit part was used; otherwise it was wood.

HYBRID MODELLING

137

The hull is now completely planked. Careful fitting ensured the plastic parts joined seamlessly to the new wooden hull. Some of the kit parts have been pre-painted (*eg*, inner bulwarks) to get a sharp paint line between the plastic and wooden pieces. The wooden parts will receive an oil finish (*eg*, tung oil) to bring out the beauty of the wood. In hybrid modelling, it is important to use a fine grained wood for the hull planking. Hardwoods such as pear have very little visible grain and this is in keeping with the scale of the model.

138

The hull is nearing completion. The wood has been sanded smooth and finished in oil. The plastic parts were painted blue and gilded with gold paint. A wash of dark brown was applied over the gilding and blue ground to create depth and remove any toy-like appearance of the bright colours. The wood grain on the underwater hull was sealed and painted off-white to represent *white stuff*, a mixture of whale oil, rosin and sulphur used to discourage ship-worm (*teredo navalis*) from boring into and damaging the hull. The mixture was likely light in colour but hardly the bright white most modellers and artists alike have portrayed.

139

The finished head. The kit parts and wooden hull are married perfectly.

140

The completed quarter galleries. The model was painted following the kit instructions. The effectiveness of real wood side planking is evident. Nothing looks more like wood than wood.

141

The masts and yards are a mix of the kit and scratch-built items made from wooden dowel for strength.

142

A view of the deck. The plastic deck received a laser cut wooden overlay from HiSModel. This veneer overlay is fully detailed with planking, butts and trennal detail. Rigging was a mixture of kit-supplied blocks and deadeyes supplemented with some purchased wooden blocks. Deck fittings like the bitts, boats, guns and carriages are all from the kit.

Rigging began with the shrouds as shown, but be aware that the rigging practices of Le Soleil Royal's time do differ significantly from those of ships of the eighteenth century and should be researched and incorporated into your model for authenticity. The new instructions for Heller's kit provides more accurate period rigging details, and R C Anderson's book *The Rigging of Ships in the Days of the Spritsail Topmast, 1600–1720* (reprinted 1994) would be a key reference for rigging during this period. The ratlines were eventually tied to the shrouds using line whose thickness and weight was in scale with the model, with the kit supplied thread discarded.

HYBRID MODELLING

143
The completed model on a custom-made display plinth. The flags and banners are printed paper items from the kit. Even if built out-of-the-box, the Heller kit is a magnificent sight that would grace any collection.

144
A stern view of the completed model. The new hull lines are far more convincing than the Tanneron museum model.

7: Modelling Realism
David Dovidio's *Vasa*
Airfix 1/144 Scale

Sweden's *Vasa* is one of the best preserved ships in the world, and along with the *Mary Rose* has taught us a great deal about how ships were built, crewed, sailed during their respective eras. The *Vasa* was built about a hundred years after *Mary Rose*, during the reign of King Gustav Adolphus and like *Le Soleil Royal* was a magnificently decorated ship that was to proclaim Sweden's emergence as a major European power. Sadly, she tragically sank during her maiden voyage on 10 August 1628. On that day, a gust of wind caught her as she sailed out of harbour, causing her to heel sharply to port. Her lower gun ports were open to fire a salute and water rushed into the ship. The ship did not have enough of a righting moment and could not recover from the heel quickly enough. At that time, the principles of hydrodynamics and stability were not well understood and erroneously many of her heaviest guns were placed on the upper decks instead of the lower decks to provide stability. Shipwrightry at that time was very much a trial and error affair, and it was likely thought that putting the biggest guns higher up in the hull would be more effective at hitting the enemy ship's side than if they were situated lower in the hull. It seems that firepower was prioritised over seaworthiness. *Vasa* was raised in 1961 after 333 years underwater. Strewn in and around her hull were thousands of artefacts, including clothing, weapons, cannons, tools, coins, cutlery, food, drink and six of the ten sails, not to mention the remains of at least fifteen people. *Vasa* can be visited at the *Vasamuseet* in Stockholm and the ship has become a symbol of Sweden.

The Airfix kit has been around since 1971 and has remained a perennial favourite with modellers. The kit's design was based on the raised ship then housed at the temporary *Wasavarvet* (The Vasa Shipyard) whilst preservation and research was underway. A great deal has been learned about the ship since the 1970s and the kit now differs in a number of details from the ship as reconstructed today. The kit mouldings are missing a number of timber and planking details likely due to the fact that they had rotted away after 300 plus years underwater, with evidence of them only discovered after extensive research. Another major difference is how the ship was painted. The Airfix kit (indeed all kits of the *Vasa* in plastic and wood alike) have tended to show all of her carvings gilded and mounted on blue painted hull sides. Research on the actual ship now suggests that the ground was red, and the carvings were all fully painted in bright colours.

Michael Dovidio of San Marcos, California, USA decided to build Airfix's *Vasa* kit, but unlike most modellers who would build the ship as she looked on that fateful day, he decided to build her as she now appears in the *Vasamuseet* with no colour at all, but just the bare preserved wooden timbers of the salvaged vessel. This presents the interesting modelling challenge of how to paint an entire plastic kit in an authentic wood finish. The kit was modified to include the missing timbers from styrene plastic strip, brass rod, and sculptures hand-made from heat polymerised clay (*eg*, Sculpy or Fimo). In striving for accuracy and realism, Mr Dovidio has undertaken to rig the model as accurately as possible, using period correct scale wooden blocks and *hearts* (early

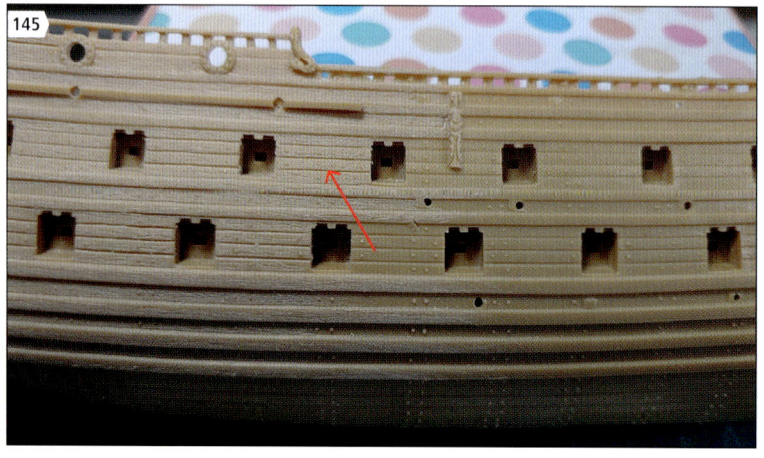

145

The Airfix hull mouldings carry some fine detail. The gun ports are moulded as deep sockets into which dummy cannon barrels are fixed. These gun ports will be opened up later because no guns are present as she appears in the *Vasamuseet*. To better capture the appearance of wooden hull timbers, the hull sides were carefully textured with additional woodgrain effects by rubbing coarse sandpaper along the length of the hull (red arrow). (*All photographs in this chapter used with permission from Michael Dovidio*)

MODELLING REALISM

versions of deadeyes), all roved using actual scale rope woven from fine threads to take advantage of all that *Vasa* has taught us about seventeenth-century rigging.

146

To the left is the original kit hull and to the right is the kit hull with additional wood graining effects and missing bow timber details added from styrene strip.

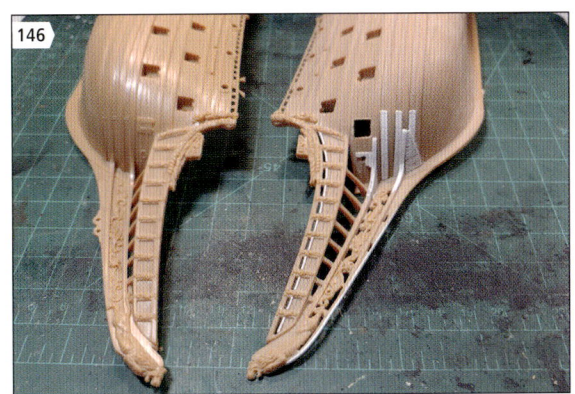

147

The kit's quarter galleries are too short in length and moulded solid. To better match the actual ship, the areas between the vertical carvings were opened up with drill, craft knife, and files. The open spaces between the vertical carvings were filled with strips of paper. The strips overlap each other's top edge to reproduce the clinker planking seen on the ship. The two domed sections were cut away and moved forward and aft to lengthen the galleries. Removal and relocation of the domes left gaps (red arrows) that were filled with paper strip to represent the missing clinker planking. The forward dome is shown complete with the new planking and a coat of light tan paint applied as a primer to check the work.

148

The modified side galleries are shown in a coat of tan paint as a primer. When the stern transom piece was fitted to the hull a space was left between the aft dome and the stern caused by correcting the shape of the side galleries and some truncation of the detail in the Airfix kit. Several experiments were carried out with different carved devices to fill this empty space. Shown in the picture is a carved figure created using heat polymerised clay (red arrow) as one of these experiments. Two windows were opened up to give a sense of depth to the interior and will be glazed. The port side of the hull shows how the kit parts fit together out of the box before the side galleries were reconstructed.

149
New channels were fabricated from .040in styrene sheet, pinned and glued to the hull (red arrow). The chainplates are copper wire (24 SWG) hammered flat and a hole drilled at the top and bottom ends (blue arrow). The heart-shaped deadeye is a white metal item by Bluejacket Shipcrafters of Maine, USA. It is stropped with black thread with the thread's tail run tied to the hole drilled into the top of the chain plate. The chainplate is bent to the correct angle, fed through a slot cut in the channel and fixed to the hull with a pin made from wire.

150
The completed foremast channels. The parts are just tacked in place to test the fit. The chain plates were painted black and the pewter hearts given a wood grain paint finish.

151
The kit's inner hull sides were extensively detailed with frames and railings built from plastic strip. In preparation for rigging, pinrails were made from .030in x .100in styrene card and fitted with belaying pins. The belaying pins were made from copper wire whose end was dipped in medium viscosity CA glue and hung upside down a few minutes to form the bulbous end. The pin was then cut to length.

This photo also shows the gun ports being opened by sawing away the moulded box. Note that the box was not cut flush with the inner hull side, but enough of the box was left to give the impression of the thickness of the hull when viewed from the outside.

152

The hull was prepared for a painted woodgrain finish with a coat of Tamiya Matt Desert Yellow (XF-59) followed by a thin wash of Tamiya Matt White (XF-2) applied with a brush.

153

The next step was to add wood shade by applying a wash of Vallejo Burnt Umber (#70941) paint thinned with water. This brand of paint is formulated for brush painting and effects such as streaking and glazing. Be aware that over-thinning acrylic paints with water causes their properties to break down, leading to the paint beading up caused by the surface tension of the water. To prevent this, acrylic medium (*eg*, Vallejo Glaze Medium #70.596) can be added to thinned paints to maintain its composition for best performance.

154

Vallejo Raw Umber (Vallejo Studio No 17) acrylic paint was slightly thinned with water or acrylic medium and brushed over the hull in one direction along the length of the planking. This coat darkens the finish and it can be repeated to create different hues and depth of colour as desired.

155

The final step to the woodgrain finish is a thin wash of matt black paint to highlight and define fine details. Once the wash is dry a coat of matt varnish is sprayed over the finish to deepen the overall tone and add warmth.

156

The kit's plastic deck was sanded smooth and a wood veneer deck from HiSModel was glued in place. New deck coamings and gratings were fitted into the hatch openings. The wooden deck was later painted over to match the brown colour of the hull. The quarterdeck bulkhead is extensively detailed with clinker planking from styrene strip, new detail carvings from clay, and given a coat of tan primer to ready it for paint wood graining.

157

The hull is shown temporarily assembled to test the fit of the decks and beakhead bulkhead. The starboard hull half has been painted and the port side unpainted but primed in this photo. The roof tops over the companionway doors have received new eaves (red arrows). The kit's solid beakhead grating has been painstakingly opened up with drills and files, and space for the bowsprit gammoning (yellow arrows) framed in with styrene strip before being painted over.

MODELLING REALISM

158

A view of the stern with the rudder hung. The kit's hull is now fully assembled, painted, and mounted on a display plinth. Mr Dovidio's wood paint finish technique yields a realistic representation of *Vasa*'s preserved timbers.

159

A view of *Vasa*'s deck and gratings. The effectiveness of the woodgrain finish to replicate how she appears at the *Vasamuseet* is clear.

160

The model is rigged with scale wooden blocks from Drydock Models & Parts. In this scale they are tiny (the smallest being 1.5mm long) but give the ultimate in realism. The rigging references included Fred Hocker's (2011) two-volume set *Vasa: A Swedish Warship* and R C Anderson's (1994) *The Rigging of Ships: in the Days of the Spritsail Topmast, 1600–1720*. Sometimes the references disagree so paintings of contemporary Dutch ships painted by the Willem van de Veldes, father and son, were studied to determine the correct run of lines. The Van de Veldes' drawings and paintings were very detailed and known for their accuracy. Masts and yards were either kit items modified to resemble how *Vasa*'s appeared, or scratch-built from wood and brass where strength to take up the tension of real rigging was required. All of the blocks were prepared by staining them brown and stropping them in thread.

161

Vasa's shrouds and ratlines are being set up using scale model rope. Scale rope can be made yourself using a miniature ropewalk (there are power and hand-cranked versions available in the marketplace) or it can be purchased ready-made in many different sizes ready to use out of the packet. The scale rope provides the utmost in authenticity, and coupled with the wooden blocks makes it difficult tell this is a kit. The photo shows a paper clamp hanging from the end of some of the ropes. The weight of the paper clamp gives the line tension until it can be tied off.

162

The shrouds and ratlines of the mainmast. The use of scale rope is very effective.

MODELLING REALISM

163
The fore topmast is being rigged. The blocks shown here are 3.5mm in size. The ratlines are tied to shrouds using clove hitch knots, and not glued across the shrouds as shown in the builds of *Victory* and *Bounty*. Mr Dovidio took a great deal of care to use the correct scale size of rope to avoid any heavy handed overscale effect, and to ensure all of the knots are in scale. Every line of *Vasa* was belayed to the correct place on the pin rails. The kit's fighting top was modified by opening up the solid areas of the moulding to form actual rails around its platform.

164
The rigging of the foremast underway. Furled sails from paper have been added. A characteristic of seventeenth century rigging is the array of lines called *crowsfeet* (red arrows) shown rigged to the forestay (left), the main yard (right), and just visible to the main topsail stay (top). It takes considerable skill to tension the crowsfeet lest they pull the lines they are rigged to out of shape.

165

The furled sails were made from of fine tissue paper. Furling a full-depth model sail is very difficult and they look baggy so a furled sail is made to one-third of the size. Rigging lines are attached to the sail and it is furled by carefully scrunching up the sail up to the yard. The rigged lines help you determine where the sail should be pulled up and scrunched to get a realistic effect.

166

The finished furled sail.

167

A Swedish ensign was made by printing the image onto fine tissue paper whose edges were taped down to a sheet of regular copier paper so it would not get jammed in the printer. The image was coated with a spray of matt clear lacquer to protect the image when it was bent and folded to hang naturally.

MODELLING REALISM

168
The bowsprit rigging underway.

169
The mizzen mast rigging.

170
One can certainly appreciate Michael Dovidio's not inconsiderable skill and effort to rig *Vasa* authentically which has paid off with spectacular results. Mr Dovodio's *Vasa* remains a work in progress, and this can be followed on the modelshipworld.com website.

8: Reconstruction of a Ship
Marc LaGuardia's *Le Soleil Royal*, 1689
Heller 1/100 Scale

Research always turns up hitherto unknown facts about a ship. Louis XIV's *Le Soleil Royal* is a case in point. It has always been assumed that Heller's kit represents the ship designed by Laurent Hubac that was laid down at Brest in 1668 and launched in 1670. Likewise, it has also been generally assumed that the 1839 model by the artist Jean-Baptise Tanneron also represents this vessel (featured in Chapter 6). However, since the kit's release in 1977, modellers researching the kit have identified several differences between the kit and contemporary illustrations by the artists Jean Bérain (see Figure 171 and 172) and Piérre Vary (Figure 173). Most notable were differences in the sculpture and decoration between the kit and the illustrations. What could account for these differences?

The answer lies in the fact that there were *two* ships named *Le Soleil Royal*. The original ship as drawn by Bérain

171

Jean Bérain's drawing of the original *Soleil Royal*'s stern. Compared to the kit parts (see Chapter 6) the stern will have to be widened to allow for six windows as opposed to the kit's five. As many kit parts as possible were used, but any differences in the extensive ornamentation will have to be carved from scratch. (*Illustration from the collection of Marc LaGuardia*)

172

The kit and Jean Bérain's drawing of the bow are generally similar, but differences are found in the decoration and sculptures. (*Illustration from the collection of Marc LaGuardia*)

and Vary was Hubac's original design. This ship was destroyed while sheltering at Cherbourg in Normandy after the Battle of Barfleur. This action took place on 19 May 1692 as part of the Nine Years War in which Louis XIV was attempting to restore the Catholic James II to the throne of England and Scotland. The French under Anne-Hilarion de Costentin, Comte de Tourville had assembled a fleet of 44 ships of the line to invade England but was repulsed by an Anglo-Dutch fleet led by the English Admiral Edward Russell and the Dutch Vice-Admiral Philips van Almonde. The defeated French fleet dispersed and made for safe ports, the majority to Saint Vaast La Hougue. *Le Soleil Royal* was too severely damaged to keep up so she was beached at Cherbourg, along with *Admirable* and *Triomphant*. There they were attacked by fireships and burnt.

After her loss, King Louis XIV resolved to replace the ship. In March 1693 he changed the name of a ship under construction from *Le Foudroyant* to the new *Le Soleil Royal*. Interestingly, *Le Foudroyant*'s master carpenter was Etiénne Hubac, the son of Laurent. Etiénne Hubac, who had overseen the original *Soleil Royal*'s reconstruction in 1688/89 and had retained the moulds (frame templates) used at the time. These now influenced the design and construction of the new ship. The new ship was somewhat larger than her predecessor in length (170ft vs 164.5ft), breadth (46ft vs 44.5ft) and draft (22ft vs 21ft) but the armament was also reduced by one gun per side on the lower gun deck. Indeed, careful examination of Tanneron's model reveals that it is a hybrid of both ships, and he probably used illustrations and drawings of both ships to build his model. He was, after all,

173

Pierre Vary's colour illustration of *Le Soleil Royal*'s quarter gallery. To accommodate its greater size, the kit's hull will have to be lengthened. The ornamentation is quite different from the kit, notably the lattice work on the upper bulwarks. (*Illustration from the collection of Marc LaGuardia*)

174

The first step was drawing by hand a 5/8in wider stern using the kit stern plate as a guide. The lower hull and upper bulwarks were also drawn in to define the area for the new quarter galleries and to determine the location and dimensions of the hull ornamentation. These drawings took a considerable length of time as each element was drawn-in, tested, erased and redrawn until they flowed seamlessly into one another. (*All photos used with the permission of Marc LaGuardia*)

an artist and not a shipwright and taking such liberties is normal 'artistic licence'. His aim was clearly to illustrate the glory of France.

Marc LaGuardia of New York City, USA embarked on one of the most ambitious conversions of Heller's kit in order to recreate the original *Le Soleil Royal* at the time of her refit in 1689. His primary reference was J C Lemineur's 1996 book *Les Vaisseaux du Roi Soleil* and the basic structural changes would mean both widening and lengthening the hull and rebuilding the stern and quarter galleries. The next, and by far the greatest, challenge would be to recreate the original ship's baroque ornamentation that would have to be hand-carved from styrene plastic, supplemented with some sections in wood. It is a good thing that Mr LaGuardia is a professional woodworker with self-taught carving skills developed over years of practice. Carving techniques are beyond the scope of this book, but there are many publications and video tutorials that show the basic techniques to get you started. The techniques used to carve styrene are not different from those used in full-scale woodworking projects. Developing the skill takes practice and patience, so begin with a small decorative item for your first attempts and after a time your skills and confidence will develop faster than you think. The first item I ever carved was a snake decoration at 1/200 scale for a scratch-built model of the 1850s French naval schooner *La Jacinthe*. It took a few tries but it was not long before I had a carved stern decoration. I recall well the sense of satisfaction and accomplishment so don't be scared to try. Mr LaGuardia has kept a detailed chronicle of his work on *Le Soleil Royal* that includes several carving demonstrations on how he recreated the ship's extensive decoration. This log can be found on the Nautical Research Journal website *Model Ship World*. This chapter can only highlight some of the five years of work Mr LaGuardia put into this project, and there remains many more years of work left to do.

175

Over the course of the build the model required several drawings to determine the size and location of the new decorations to fit the available space. To assist with the design process, the manual drawings were digitised with the open source CAD software that helped speed the editing process.

176

With the preliminary design finalised, it was time to cut plastic. Modification of the hull began by framing-in the lower gunports with .125in square rod, and the upper gun ports with .100in square rod. The aft gunport has been moved forward by 1/4in and the gap filled with styrene strip. The hull below the waterline was cut away creating a model that will eventually be set into a seascape.

RECONSTRUCTION OF A SHIP

177

In order to accommodate the dimensions of the new quarter galleries, a 1/2in extension was added to the stern. The extension piece is from an unfinished kit donated by a fellow modeller. The hull sides were detailed with iron bolt details used to fasten the planking on the real ship. Correctly scaled dome-headed bolts were made by cutting 1/64in thick slices from .020in styrene rod with a single-edged razor blade. The slices are laid in a single layer on a flame proof surface (eg, a ceramic plate). The flame from a barbecue wand lighter is passed above the wafers (about ¼in) allowing the heat from the flame to 'dome' each slice. It will take some practice with the height and speed to get the plastic discs to dome properly. The bolt heads are picked up with a tip of a knife and set into a spot of glue on the hull.

178

The donor kit also provided bow sections to widen the hull sides. The planking detail has yet to be re-scribed in this photo.

179

A start was made on the new decorations with the lattice work on the upper bulwark. A critical design challenge is that the lattice should mirror the aft rake of the stern transom while maintaining a consistent spacing. The frieze lattice was painstakingly cut out of sheet styrene and the ornaments are resin castings (*eg*, Smooth-On or Aluminite) poured into a room temperature vulcanising mould (RTV) made from a hand carved master.

180
Once the frieze lattice was carved it had to be applied to the hull side so that it maintained the elegant sweep of the hull's sheer line. The frieze was cut into vertical segments along the 'crossed Xs' and fixed in a position onto the hull to ensure the correct sweep. The joints were covered-up by resin-cast quatrefoil ornaments. Note how some of the lattice has been carefully relief-carved to suggest the braiding that eliminates a flat presentation. A new upper railing was also made from styrene to match the Vary drawings.

181
Experiments for the painted wood finish was carried out on a spare hull section from another kit. Several variations were tried using different combinations of primers and paints. The colour eventually selected for the model was obtained by first priming the hull white and over that thinned coats of artist's acrylic raw sienna were applied by brush. Once the raw sienna had cured, a liberal application of Windsor & Newton Van Dyke Brown oil colour was applied with the majority wiped off the surface; the dark colour tints the surface and deepens the cracks and crevices imparting depth and highlighting the coarsely sanded woodgrained surface of the planks. This colour is called *ventre-de-biche* (belly of a doe).

182
The photo shows the colour of the wooden hull. The kit's port lids were detailed by adding plank lid-linings and cast shell and fleur-de-lis ornaments that echo those of the ornamental frieze above.

183
After the main colours of the hull halves were blocked in, the next task was to widen the stern to accommodate the sixth window. The compound curves of new stern shape were framed in with profiled bulkheads cut from styrene and glued in place.

184
The bulkheads were planked over with styrene strip in the traditional manner. A very slight bevel on both plank edges was sanded in so that the plank seams will not completely disappear under paint. A new rudder post has also been fitted.

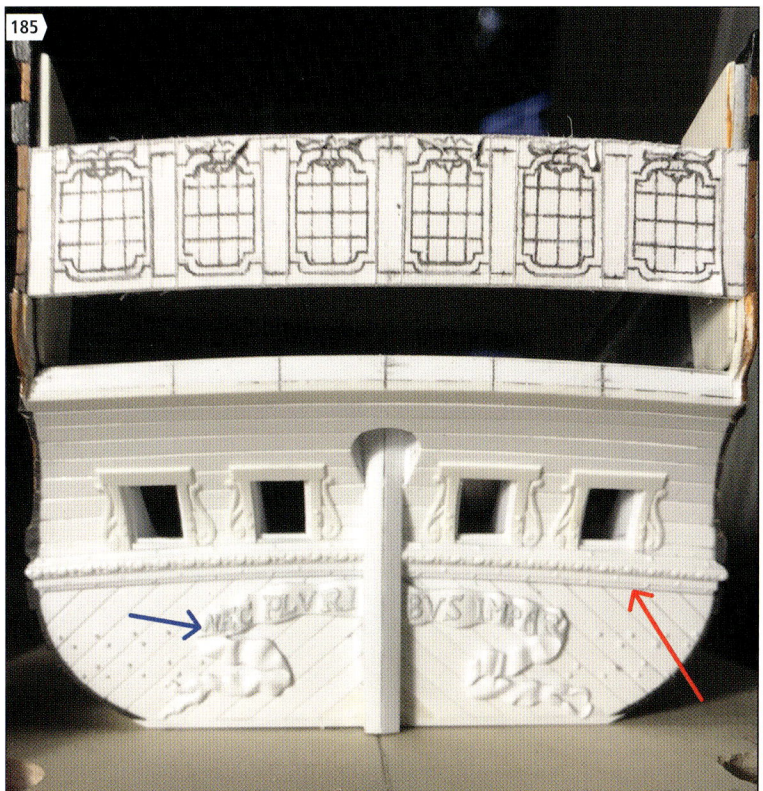

185
Attention now turned to decorating the planked transom. The first item made was a detailed moulding made up of three different sizes of styrene strip that have been edge-profiled with scrapers made from old hacksaw blades (red arrow). The outermost layer of the moulding is a half-round styrene strip that has been carved into a simplified egg and dart pattern that matches Bérain's stern drawing.

Marc LaGuardia exercised a bit of artistic licence and added the scroll to the transom (blue arrow) bearing the Latin phrase *Nec Pluribus Impar* as a reference to Louis XIV's belief that he was 'not unequal to the illumination of many suns'. Although there is no evidence that such an ornament existed on the real ship, a similar banner and motto are emblazoned on all of the ship's bronze guns. The letters are individually cut from parchment paper and stiffened with thin CA glue. The scrolls around the gunports were individually relief-carved by hand. At the top, a card template for the great cabin windows is being tested.

186

The first tier of windows has been fitted, and the completed rudder with carved dolphin is shown in place. The rudder ornament of dolphins (red arrow) was a little more artistic licence in keeping with the fact that the ship was awash in carved dolphins. The sculpture was recreated from the stern quarter drawing by the artist Pierre Puget of the newly refitted *Dauphin Royal* of 1680.

187

Windows are clear acetate sheet that was scribed with a sharp knife. Medium-grey acrylic paint was wiped into the lines to create the frames for the window panes.

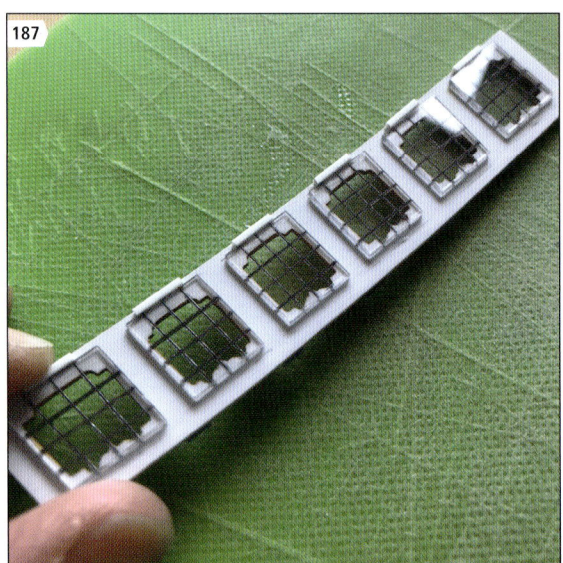

188

A card template was made for the new main deck and the shape was traced to styrene card and cut out. All of the individual deck plank lines, scarph joints, and the shift of the butts were scribed into the plastic sheet. In order to create the cambered look of the gratings, grating parts from two kits were glued together and the tops sanded to a rounded-over shape.

189

The new main deck was painted with acrylic light tan, followed by a wash of Windsor & Newton Payne's Grey which highlights the careful engraving work. Once cured, a wash of Van Dyke Brown is brushed over the top for a weathered look.

190
The quarter galleries were redrawn by hand and test-fitted to the hull. The drawing was adjusted until it fitted the space exactly at the correct angle.

191
The upper bulwarks with the top section of the hand carved quarter galleries are test-fitted to the hull. Even without paint, the beauty of Mr LaGuardia's carvings hold their own.

192
Work begins on the lower quarter galleries. The basic outline has been cut from wood stock because the undulating shapes required significant depth of material, and wood is just easier to work with for this purpose. There are three distinct levels to the lower galleries and each level was carved separately.

193
The finished starboard lower gallery piece to the left. On the right is the port side piece receiving decoration blanks cut from .020in styrene sheet. The blanks will have its detail carved in to match the other side.

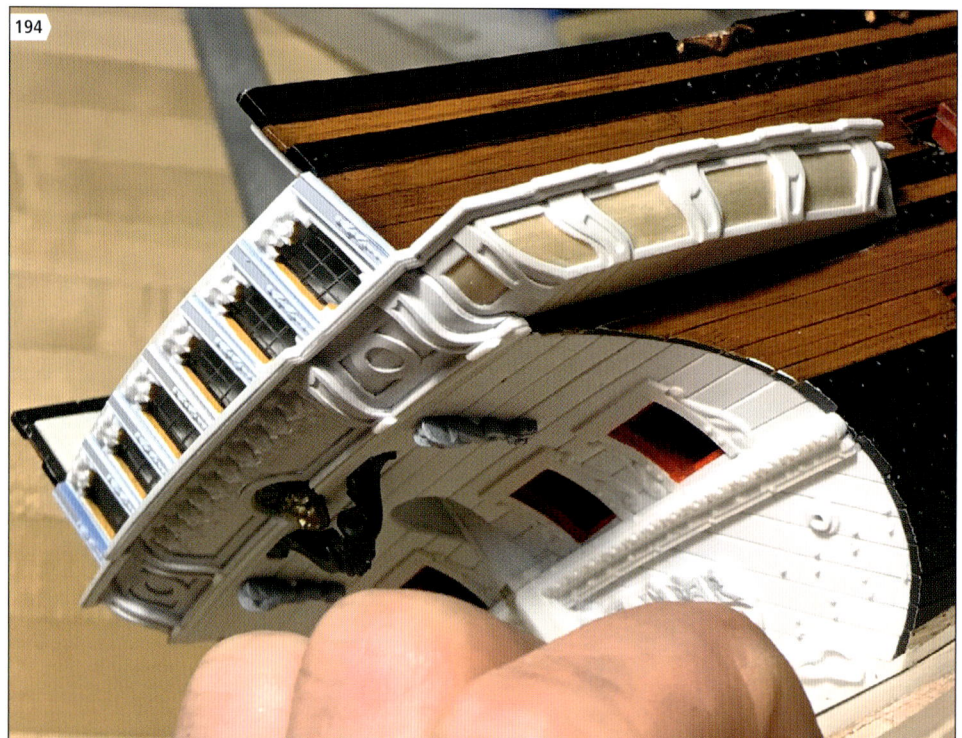

194
The next level of the lower quarter gallery is also carved from wood with styrene decoration. This part is being test-fitted to the hull to ensure that it fits neatly into the extensions of the stern counter.

195
The lower stern and quarter galleries completed, painted and added to the hull. The application of paint really brings the beauty of the carvings to new heights. *Le Soleil Royal* must have been a magnificent sight under sail.

The kit figures of the Four Seasons required styrene shims to give each figure a forward leaning posture, as well as additional height to support the balcony above. Also note that the lower windows of the quarter gallery are painted black with stylized highlights because on the real ship they are *false lights* – serving a decorative function only.

RECONSTRUCTION OF A SHIP

196

The aft section is now fully painted in all its resplendent glory. A few words about the colour scheme are necessary. Most depictions of the ship in model form and in Vary's illustration show the carvings gilt on an ultramarine background. The *Vasa* has also been depicted through the centuries in a similar scheme, but recent archaeological research has shown that not to be true. The reason is likely artistic licence where the artist wanted to show the glory and wealth of their Monarch who can afford to decorate his ships in this way. Ultramarine paint was a very expensive pigment made from crushed lapis lazuli, whose cost rivalled gold leaf, and provides an explanation for why the actual ship was not painted in this way. The same arguments can be made for *Le Soleil Royal*, hence the bright colour scheme chosen. Consequently, gold and ultramarine were typically reserved for ornamental areas that made direct reference to the Crown, like the Arms of France, for example.

196

197

With the side galleries in place, attention turned to completing the stern galleries. The six windows were framed and the camber was slightly increased to set the stage for the large allegorical carving of Apollo, as the Sun King rides his horse-drawn chariot across the sky (red arrow).

198

The windows are fitted and all of the sculptures and decorations painted.

199

Finally, the magnificent stern is finished. The area occupied by the carving of Apollo and his chariot is significantly smaller than the space on the Tanneron and Heller models, but this is consistent with the French imperative to reduce top-hamper, and thus improve the weatherliness and manoeuvrability of their largest vessels.

199

112　　　　　　　　　　　　　　　　　　　　　　　　　　　　　　SAILING SHIPS FROM PLASTIC KITS

RECONSTRUCTION OF A SHIP   113

200
Attention now turned to the stem. The kit's figurehead was used without modification but the kit's detailed *trailboard* that fits between the upper and lower knees of the head was fretted out of the solid kit moulding (yellow arrow). A realistic timbering layout was scribed into the cutwater, tapering its forward edge, adding bolt heads, and fashioning a block for the foresail tack (red arrow). New headrails were cut from 1/16in styrene sheet with the horsehead and rosette relief detail carved-in, embellished with kit parts in grey plastic.

201
The completed headrails are fully painted and the cathead support timbers have been fixed in place. The bow also features winged female caryatids (blue arrow), and just visible within the headrails is the new headrail support and grating structure (red arrow) that will be properly cambered, athwartships, while also following the upward sweep of the headrails. Scale rope has been fitted into the hawse holes. In this view the scratch-built channels have been fixed to the hull with supporting knees.

202
Le Soleil Royal inches towards completion, and in this photo a start has been made on the masts. There remains the making and fitting of the upper decks and the armament. Then there are also the yards, rigging, and sails to build and fit. A very detailed step-by-step account of Mr LaGuardia's progress can be found on the modelshipworld.com website, along with other builders' models of the Heller kit that offer many different techniques and ideas to study.

Ambitious projects like this conversion of *Le Soleil Royal* are made possible by the generous assistance of the international modelmaking community, and Mr LaGuardia would like to acknowledge Messrs Michel Saunier, Guy Maher, and Marc Yeu who freely shared their techniques and research findings to make the project possible.

9: Life at Sea
Daniel Fischer's *HMS Victory* Heller 1/100 Scale

The American philosopher Ralph Waldo Emerson is reputed to have written 'It's not the destination, it's the journey' and this sums up the approach to modelling Heller's 1/100 *HMS Victory* taken by Daniel Fischer of Ludwigsburg, Germany. *Victory* is a ship with over 250 years of history, and during her life her appearance has changed dramatically. She was not always the black and yellow ship we see in Portsmouth today, but when she was launched she had an open stern and a payed finish at her commissioning in 1776. In 1783 her sides were painted for the first time in yellow ochre with black wales, and in 1803 she received a paint scheme of ochre with black stripes, and it remains unknown exactly when the outside of her port lids were painted black to create the famous 'Nelson Chequer'. In 1816 her hull was radically altered and received a round bow; her hull was painted black with white stripes carried out to the extremities of the hull, a scheme she was to carry for more than a hundred years. She was restored to her familiar ochre and black in the 1920s and the round bow was removed when she was converted into a museum ship. Since then she has been restored and repainted several times, and with each restoration brought variations in details and her rigging as little known facts are revealed by research. The ship is rather like a Supermarine Spitfire: there are so many variations of the airframe and squadron markings a modeller can spend a lifetime building nothing but Spitfires.

Victory's continued existence has meant that a lot of research on every timber and fitting has been conducted over many years by scholars, conservation experts, and model makers – so much so that it is now possible to determine with certainty the exact shade of 'yellow ochre' used to paint her hull in 1805, so as she appeared at the Battle of Trafalgar.

203

This display shows how the large anchors were handled. Shifting the anchors required a huge number of men and ropes to handle its massive weight. The kit anchors were replaced with scratch-built items made to contemporary plans for a more true-scale appearance. The catheads are 3D printed items designed from plans to better catch their shape, heft, and detail. (*All photos in this chapter used with the permission of Daniel Fischer*)

Dockyard records and archaeological methods that literally peeled back the layers of paint was matched to the recorded dates it was applied. Research findings such as this emerge all the time as the ship is studied. Presently, *Victory* is under restoration again and for this effort all of her planking is being removed, exposing the frames. It is exciting to think about what next will be found. This is all very inspiring to a modeller and using Heller's kit as a basis Mr Fischer has built and rebuilt *Victory* at least a dozen times since 2002. Inspired by the latest research, he has embarked on a journey to build several in-depth studies of *Victory* but in Ralph Waldo Emerson's terms, having a completed model is not his goal, but rather it is a journey of *discovery*.

This journey has taken Mr Fischer down many roads, and each road is not a dead end, but rather is another path to understanding *Victory*. His early work was to create an accurate appearance of the ship, and to this end he has designed a large set of photoetched parts to refine the kit parts, and a series of resin parts to accurately detail the ship at different times in her long career. For example, *Victory*'s stern sports the three Prince of Wales feathers. However, these feathers were taken from HMS *Prince* and only mounted on *Victory* in 1837, many years after the Battle of Trafalgar. What did she sport at this famous battle? Historians and scholars suggest it was a bust of George III, so a set of possible busts of George III was designed to fit the kit. Others suggest the ship had the crest supported by a lion and unicorn. Similarly, Heller's kit hull follows a block model in the National Maritime Museum when the ship had no side entry ports. These ports were cut into one or both sides of the hull at different times according to records, and both are now topped with a set of ornate eaves. Mr Fischer created these parts so any variation of the ship at any time can be modelled.

When you visit *Victory* in Portsmouth, the guides will give you a graphic description of what life was like on board her. Mr Fischer decided that to really know the ship, he had to model that shipboard life. To this end, a series of vignettes were created displaying life on the mess deck, manning the main capstan to haul in the anchor, relieving nature's call at the head, and different stages of the drill on the great guns. These vignettes were all created within the Heller hull mouldings and with the personnel converted from model railway figures. *Victory*'s ever-changing appearance has been recreated in a series of tableaus that have taken sections of the Heller kit and modified them to her appearance when launched, at Trafalgar, in the 1920s and now as a tourist attraction in Portsmouth. Each of the tableaus is complete with figures to give a sense of scale, life, but most importantly the 'feel' of the time period being portrayed. What is remarkable about Mr Fischer's models is not just the technical skills, but his imagination and desire to get to know the ship inside and out, and to be one of those figures at the mess table.

204

The evolution of a cathead. To the left the cathead's body was made of styrene using the kit parts for the decoration, sheaves, and cleat. This was used to help design the 3D part shown in the middle. This version was a little undernourished and a third version (right) was designed to its true dimensions and printed out.

205

A scene on board the ship showing how the anchor cable was attached to the capstan's traveller cable with short lengths of rope called 'nippers' that hauled up the anchor. The ship's boys were often in charge of the tying on and removing nippers and is the origin of the term 'little nipper' when referring to some bright young lad. All of the items in this photo, apart from the guns, are made from scratch. The guns received a PE Royal Cypher. The care taken in planking the deck can be seen at the top right of the photo. The deck planks were laid individually and properly joggled into the waterway.

206

The kit beakhead bulkhead (bottom) and a modified version showing open doors and ports. The door and port openings were lined with sills and jambs for the utmost in detail. The kit's top rail was modified by drilling out the solid detail. After painting, all of the details were outlined with black ink to add some depth.

207

Gratings and hatches allowed air and light into the ship. These were all made from scratch from scale wooden battens (.4mm thick by 2mm wide) cut and slotted from wood salvaged from an old shelf.

208

A view of the inside of the stern of the ship under construction. The basic rudder head assembly is tested for fit and a seated figure is used to check the scale. The structure and layout of this area is true to contemporary sources, but differs from *Victory* today because of her many modifications and rebuilding over the past 250 years.

LIFE AT SEA

209

The ship's cooking range, often referred to as a 'Brodie Stove' after the stove's manufacturer, before painting. The stove was made up from styrene sheet, blocks of polyurethane resin, wood, and brass. It is really a matter of using whatever material works best.

210

The capstan went through two decks and this photo shows the lower portion with the *traveller cable*. The capstans were scratch-built from contemporary plans. The palls (or pawls) to stop the anchor from turning backwards are workable and could be flipped into the opposite direction for when the traveller cable is moving the other way.

211

Victory's chain plates and deadeyes accurately modelled. These are made of meticulously formed brass wire. The advantage of brass wire is that the parts look like they were forged from iron bar as on the real ship. Mr Fischer has designed alternative PE parts that can be used, as does ScaleWarship. The PE parts do greatly simplify the construction of these details and provides delicacy.

212

The kit's underwater hull was covered with self-adhesive copper foil that was detailed with nail head detail with this homemade ponce wheel.

213

The coppered rudder and hull with detailed copper plates. Each plate was laid in the correct pattern and overlaps. The PE draft marks really add authenticity. The gap between the stern post and rudder on *Victory* today is much larger than that shown on contemporary draughts so the gap was modelled much smaller.

214

The copper plates were given a patina with paint. Note that there are three very distinct colours. The top row at the waterline has a greenish tinge that happens when copper is in contact with salt water and a great deal of oxygen. The next two strakes are brighter because 1) this area is under water but has a greater oxygenation being close to the surface and, 2) is subject to greater abrasion from the water molecules in the disturbed water when the ship is under way. The remainder of the hull is under water and the dark colour represents prolong immersion in water with less oxygen.

LIFE AT SEA

215

The method used to make sails for *Victory* is shown here. The sail is a fine cloth to which strips of transparent paper used to restore old antique books is placed on both sides. The paper strips are coated with a heat-activated glue and are ironed on to the cloth with a domestic iron. This work is carried out over a light box to ensure the strips are perfectly aligned on both sides of the cloth. There are no imitation seams sewn or painted on the sails, instead relying on the transparency of the material and the opaqueness of the double layer of the paper to give the effect. In this photo a light is being shone through the sail to ensure that the translucent nature of the cloth is retained to replicate the appearance of a real sail in bright sunlight. The reef bands have holes punched into the paper for reef points.

216

A completed sail showings seams, boltropes and cringles (the metal or wooden loops used to bend a sail to the stay). With this method the sails can be brought into almost all natural positions, and even furled sails can look very realistic.

217

The kit's bowsprit under construction and detailed with styrene strip. The kit's iron hoops were sanded off and replaced with strip styrene. The kit parts are moulded in two hollow halves allowing them to be strengthened by gluing in a brass rod when assembling the halves.

218

The painted bowsprit showing the cap, martingale and jackstaff made from wood to take the strain of rigging.

219

The kit's masts under construction with additional detailing from plastic strip. The moulded iron hoops were sanded off and replaced with strip styrene. The masthead was replaced with a scratch-built head to capture its true tapered form.

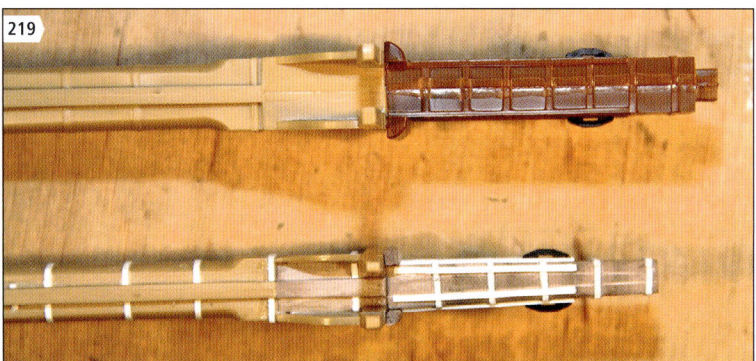

220

The completed masts and tops. The figure gives you a good idea of how massive they are. The design of the mast tops parts followed the 1802 Admiralty's instruction to fit the tops in two halves around the mast. Now that is authenticity!

LIFE AT SEA

222

The main yard with its set of *jeer blocks* (double or treble sheaved blocks used to hoist a yard), *sheet blocks* (used to help form the shape of a sail to capture the wind. Sails work like a crude aerofoil, and the shape of the sail determines the high and low pressure areas that provides motive power) and *clew garnets* (a rope attaching the clew – or lower corner of a sail) used to furl and unfurl the sail. The blocks shown are 3D printed and painted.

221

A completed mast top with lantern and railing. The yard is held to the mast with a chain sling. The block and truss shown here is the topmast lift and serves to bring up and down the topmast.

223

Victory's massive head gear. It has all the bobstays and shrouds with printed hearts and hand-made scale rope rigged in the appropriate diameters.

224

The interior of the quarter galleries under construction. These galleries carried the Officers' toilets. Note the officer's uniform coat hanging on the ship's side and the classic lantern. The decoration on the quarter galleries was sculpted with A+B putty (*eg*, Magic Sculpt, Greenstuff).

226

A tableau of *Victory*'s appearance in 1776. The crew are busy swabbing and holystoning the deck under the gaze of a Royal Marine who is probably thankful this was not his job. The ship's guns also sport white canvas covers to keep the water out.

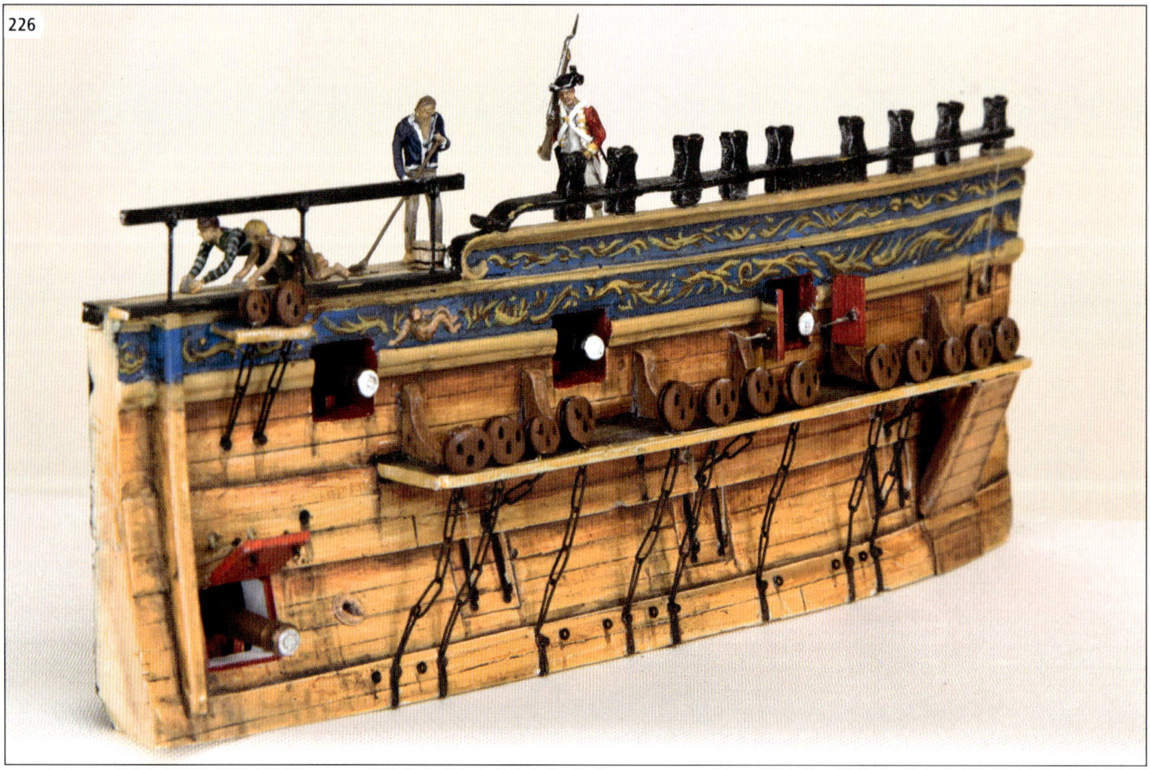

225

Victory's completed stern ready for installation. Note that the Prince of Wales feathers have been replaced with a Crown which she could have carried at Trafalgar, though that is not known for certain. This part is the original kit part where the window frames have been thinned with needle files to about a third of their moulded size. Mr Fischer and ScaleWarships offer PE accessory sets that capture the delicacy of the frames without the tedious filing.

227

A tableau of *Victory* after her 1803 refit. The crew are hanging off the side making good the damage caused during the great chase of the French fleet into the Caribbean before the Battle of Trafalgar. The matelots are swabbing on colours of the pink version of yellow ochre that was known to be on *Victory* at this time. The gun port lids have yet to be painted black to create the famous 'Nelson Chequer'. The Marine's headdress of the time was accurately modelled as a round top hat that replaced the tricorne. He is probably thinking 'better those fellows than me!'

228

Victory as store ship in Portsmouth Harbour in 1913. There is an existing contemporary picture (from the collection of Daniel Fischer) of a sailor on submarine *C34* lashed alongside *Victory* and we see the officer on the model 'taking' this picture. The dress of the sailors for this time period has been accurately modelled.

229

Victory as repainted in 2015 to the terracotta scheme (pink version of the yellow ochre). There are no sailors on this model, just portraits of Mr Fischer and his extremely patient and understanding wife.

230

Model railway figures were cut up and repositioned to have them man the main capstan.

231

The completed capstan and crew of seamen and Marines fitted to *Victory*'s hull. All that is missing is the fiddler who sat on the capstan playing a tune to set the rhythm. You can almost hear the stamping feet and grunting of the crew.

LIFE AT SEA 125

232

One of the ship's guns on the main gun deck in action. Imagine the smoke and noise in this confined space during a battle.

233

A vignette of *Victory*'s foremast with the ship at sea. The attention to detail such as the weathering and paint chipping really conveys the feeling of a working warship. Every detail is convincingly modelled, from the tompions in the guns, to the way the anchor is slung with chains and rope, where each of the detailed figures is placed from the leadsman, to those climbing the rigging, all help to recreate a realistic scene.

234

The conviviality of the mess deck. The Marines had their messes at the end of the gun deck that formed a barrier between the officer and warrant officer quarters. Here they slept and ate as a separate group. Note the square trays that held their tin plates, the origin of the term 'having a square meal' that is still used today.

235

The Gunroom and the Warrant Officer's mess. The youthful exuberance of the midshipmen who were boys and teenagers (top left) is evident. Once again Marines are stationed outside their mess, to separate the 'people' from the officers. Screens are erected between guns to create 'private' spaces for each officer. *Victory*'s gunner, Mr Rivers, lived in the rearmost of these small cabins for 22 years.

236

Answering Nature's Call at the Head. In good weather this must have been a very busy place on the ship with only six seats of ease for about 800 sailors. There is no privacy on a ship, which epitomises life at sea.

Daniel Fischer's unique approach to modelling plastic sailing ships can be followed on the modelshipworld.com website.

Resources

This book has shown that a plastic model ship kit can be turned into a beautiful model of a sailing ship with as little as some minor modifications, careful assembly, and painting out-of-the-box that captures the spirit of ship and the sea. A little or a lot more work can turn the kit into an exacting replica of the ship at a particular period in her history. Whatever approach you choose or technique you try, the success of it lies with good research and understanding the parts of a ship, and the mechanics of the modeller's technique.

There are hundreds of books on sailing ships suitable for plastic modellers, and locating and accessing them can be quite a chore. The following are those listed in the text and a few other useful general references. Some of the books were originally published as far back as the 1930s but have been reprinted several times and are readily available for purchase or often can be found in a public library.

Selected References

Anderson, R C (1994). *The Rigging of Ships in the Days of the Spritsail Topmast, 1600–1720.* New York: Dover Publications.
Boudriot, J (1986). *The Seventy Four Gun Ship,* Vols I-IV, Paris: ANCRE.
Goodwin, P (1987). *The Construction and Fitting of the English Man of War 1650–1850.* London & Annapolis: Conway Maritime Press & Naval Institute Press.
Harland, J H (1984). *Seamanship in the Age of Sail: An Account of the Shiphandling of the Sailing Man-of-War, 1600–1860, Based on Contemporary Sources.* London: Conway Maritime Press.
Hackney, N C L (1970). *HMS Victory.* London: Patrick Stephens.
Jang, K L (2022). *Ship Models from the Age of Sail: Building and enhancing commercial kits.* Barnsley: Seaforth Publishing.
Jang, K L (2022). *ShipCraft 29. Victory: 100-gun First Rate 1765.* Barnsley: Seaforth Publishing.
Jang, K L (2023). *ShipCraft 30. Bounty: HM Armed Vessel 1787.* Barnsley: Seaforth Publishing.
Lavery, B (1987). *The Arming and Fitting of English Ships of War 1600–1815,* London & Annapolis: Conway Maritime Press & Naval Institute Press.
Lees, J, (1990). *The Masting and Rigging of English Ships of War 1625–1860.* London & Annapolis: Conway Maritime Press & Naval Institute Press.
Lemineur, J C (1996). *Les Vaisseaux du Roi Soleil.* Nice: Omega.
Longridge, C N (1959). *The Cutty Sark.* New York: Edward Sweetman.
Longridge, C N (1987). *The Anatomy of Nelson's Ships* (10th impression). Annapolis: Naval Institute Press.
Marquardt, K H (1990). *Eighteenth-Century Rigs & Rigging.* London: Conway Maritime Press.
McKay, J (1987). *Anatomy of the Ship. The 100-gun Ship Victory.* London: Conway Maritime Press.
McKay, J (1989). *Anatomy of the Ship. The Armed Transport Bounty.* London: Conway Maritime Press.
McNarry, D (1975). *Ship Models in Miniature.* New York: David & Charles.
Petersson, L. (2000). *Rigging Period Ship Models.* London: Chatham Publishing.
Petersson, L. (2007). *Rigging Fore and Aft Period Craft.* London: Chatham Publishing.
Reed, P (2000). *Modelling Sailing Men-of-War.* London & Annapolis: Chatham Publishing & Naval Institute Press.
Reed, P (2007). *Period Ship Modelmaking.* London & Annapolis: Chatham Publishing & Naval Institute Press.
Underhill, H A (1938). *Sailing Ship Rigs & Rigging.* Glasgow: Brown, Son, & Ferguson.
Underhill, H A (1994). *Masting and Rigging the Clipper Ship and Ocean Carrier.* Glasgow: Brown, Son, & Ferguson.

Research

Museums and organisations around the world have searchable indices and collections for basic research such as Royal Museums Greenwich (the National Maritime Museum) archives. For French vessels the Association des Amis du Musée de la Marine is highly recommended.

For the plastic (and wooden) ship modeller an organisation dedicated to nautical research and model making is the Nautical Research Guild (NRG), 237 South Lincoln Street, Westmont IL, 60559-1917. Membership in this organisation comes with a subscription to their journal that features articles on research, history, and plenty of model making. The NRG hosts a huge website called Model Ship World. This website has kit reviews and hundreds of build logs for model ships built in all materials.

Accessories and Kits

Plastic kits of sailing ships can be found at model shops or from online retailers. Below is a list of suppliers that offer accessories, such as real wooden blocks, etched brass detail sets, rigging materials, paints, or specialised kits mentioned in the text. Many of the offerings were designed for wooden kits but, as we have seen here, can be used on any model regardless of material.

BECC Model Supplies
http://www.becc.co.uk/
A large range of cloth flags, pennants, and vinyl lettering perfect for model sailing ships in a large number of scales. They are well printed and accurate.

Bluejacket Shipcrafters
http://www.bluejacketinc.com/
A wide range of highly detailed ship fittings all cast in pewter suitable for any kit, plastic or wood.

CAF Models
https://cafmodel.com/
This wooden ship manufacturer produces a high quality range of fittings easily adapted for plastic kits and has introduced a range of small-scale 3D resin cast and printed kits with wooden deck overlays.

Dafinsmus
http://www.dafinismus.de/index_en.html
If you are building Heller's 1/100 HMS *Victory*, a full range of etched and 3D printed parts are available that cover all details of the ship and rigging. They are designed by Mr Daniel Fischer whose models of *Victory* in Chapter 9 features many of these parts.

Domanoff Workshop
https://shipworkshop.com
Tools such as ropewalks to make your own scale rope.

Drydock Models and Parts
https://drydockmodelsandparts.com/
A wide range of sailing ship fittings, guns and rigging requirements, plus high quality wooden ship kits.

HiSModel
https://www.hismodel.com/
This online shop specialises in accessories specifically designed for plastic kits of sailing ship. They offer etched details, turned cannon, accurate cloth flags and sails, wooden rigging blocks, all tailored for specific kits. Also offered are wooden decks for almost all of the popular kits on the market. The owner is a dedicated plastic modeller and has designed his product offerings with plastic modellers specifically in mind.

Langton Miniatures
http://www.rodlangton.com/napoleonic/list.htm
A wide range of 1/1200 scale Napoleonic era vessels in white metal, primarily intended for wargaming. Also more detailed 1/300 scale multi-media kits of sailing warships.

Model Expo
https://modelexpo-online.com/
A huge range of fittings for models of sailing ships in all scales. The fittings are offered in wood, white metal, and brass. A wide range of rigging cords, paints, reference books and plastic and well-designed quality wooden kits.

Scale Warship
https://scalewarship.com/
A growing selection of photo-etch and resin accessories aimed at plastic models of sailing ships.

Syren Ship Model Company
https://syrenshipmodelcompany.com/
Kits and semi-kits and a wide array of tools and supplies, including rigging blocks and fittings, and ropewalk to make your own scale rope.

Henry Turner
https://www.turnerminiatures.co.uk/
A wide range of exquisitely detailed printable 1/700 ships.

Vanguard Models
https://vanguardmodels.co.uk/
Growing range of accessories and fittings such as wooden rigging blocks and etched brass fittings. The wooden kits in this range are modeller friendly and authentically detailed and an ideal introduction to wooden ship models for those used to plastic kits.

General Model Shops

Cornwall Model Boats
https://www.cornwallmodelboats.co.uk/index.html
The Admiralty Paint range suitable for sailing ships.

Hannants
https://www.hannants.co.uk/